APPLYING THE SYSTEMS APPROACH TO URBAN DEVELOPMENT

COMMUNITY DEVELOPMENT SERIES

SERIES EDITOR: Richard P. Dober

COMMUNITY DEVELOPMENT SERIES

APPLYING THE SYSTEMS APPROACH TO URBAN DEVELOPMENT

JACK W. LAPATRA

Associate Professor
Department of Electrical Engineering, College of Engineering
Department of Community Health, School of Medicine
University of California, Davis

Dowden, Hutchinson & Ross, Inc.

To Scott and Bill

Library of Congress Cataloging in Publication Data

LaPatra, Jack W   1927-
    Applying the systems approach to urban development.

    (Community development series, v. 5)
    Bibliography: p.
    1. Cities and towns. 2. System analysis.
I. Title.
HT153.L28   309.2'62   73-11942
ISBN 0-87933-042-2

Copyright © 1973 by Dowden, Hutchinson & Ross, Inc.
Community Development Series, Volume 5
Library of Congress Catalog Card Number: 73-11942
ISBN: 0-87933-042-2

Manufactured in the United States of America.

74 75 76   5 4 3 2

Exclusive distributor outside the United States and
Canada:
John Wiley & Sons, Inc.

Preface

More and more people are becoming interested in applying systems analysis to problems in the public sector. This book is aimed at contributing to the understanding of phenomena associated with the urban setting, and to facilitating problem solving within the urban setting. The level of the exposition is carefully controlled so that the book can be read by a wide variety of professionals. Results and examples are presented for many aspects of urban development, and although these will not convert the reader into a systems analyst, the material indicates the capability and promise of the systems approach. The Bibliography at the end of the book provides a guide to further reading.

There is no nontrivial urban problem today that can be dealt with by a single discipline or group of professionals. The systems approach provides the vocabulary and viewpoint to bring together the necessary skills and knowledge in a problem-solving collaboration. In addition, systems methodology has a wide variety of tools available for the actual problem solving. In almost all cases, the urban problems considered do not yet have their own highly developed methodologies and literatures.

In the first part of the book an introduction to the systems approach is given. This includes a history of the systems approach, an exposition of its techniques and applications, and an extensive review of the philosophy and limitations of the problem-solving methodology. Following this is a presentation of the tools of systems methodology, with emphasis on gaming, planning–programming–budgeting, cost effectiveness, decision making, sensitivity analysis, linear programming, and simulation. Examples of each are given.

In the second part of the book, urban development models are considered for planning, transportation, and regional analysis. There is a description of the urban setting, a review of urban theory, and a presentation of the conceptual spectrum of extant urban models, urban performance measures, and methods of prediction. A survey of the modeling strategies for transportation systems is given in addition to discussions of limitations of right-of-way planning, multimodel systems, and air-routes planning and control. The macromodeling associated with urban regions is also examined.

The modeling and analysis of urban subsystems in Part 3 includes the topics of community health, justice systems, urban protective services, education, and welfare.

The last part of the book looks at a variety of urban subsystems whose analysis is only preliminary. In addition, urban modeling, theory development, and prediction from the viewpoint of Jay Forrester is studied as an indication of future uses and promise of the systems approach.

This book is dedicated to the idea that every day hundreds of decisions are being made in the public sector, and, by any standard, many of them are very poor decisions. The systems approach offers our public decision makers another information source in the hope that the quality of public-sector understanding and decision making will increase.

Special acknowledgment and thanks are due Gary Konas, who collected the background material, and Dave Gillooly, who prepared the illustrations.

J. W. LaPatra
Davis, California
April 1973

Contents

Part I
Introduction to the
Systems Approach

CHAPTER 1

Overview of the Systems Approach

To begin to discuss systems methodology and its value in problem solving requires a systematic approach because of the wide variety of topics that must be considered. In this opening chapter we shall examine the systems approach, its history, limitations, and promise.

First, of course, what is the systems approach? The systems approach is the utilization of a set of methods, techniques, and intellectual tools, collectively known as systems analysis, for complex problem solving. It would be extremely misleading to define the systems approach, as is often done, in terms of a series of steps or operations to analyze something that exists, or to design something new. The versatility of systems methodology is displayed only when there is consideration of context or purpose. To establish or adjust the complicated scheduling methods of an outpatient clinic providing a multitude of health services by merely viewing it as a complex scheduling problem is as meaningful as an evaluation of open-heart surgery by doing a time-and-motion study. The emphasis on a definition of the systems approach must be in terms of the problems to which it is most applicable, and what one might hope for as a result of applying the methodology.

The application of the tools of systems analysis grew slowly as an unstructured and sometimes illogical way to meet the demands for solutions to important problems. But, before we go further, we must give some thought to a second basic point: What is a system? For our purposes we shall think of a system in the most general way as a collection of entities interrelated in a specific way to accomplish a particular objective. The members of the collection are usually termed "subsystems." Clearly, with this definition we can include almost anything.

The bicyclist pedaling down a city street is a system with two obvious large subsystems: the person and the bicycle. Smaller subsystems include the wheels, the operator's eyes, and his nervous subsystem. This viewpoint indicates that any system we choose to consider must itself be a collection of objects in a hierarchy. The bicyclist is a member of the city bicycle traffic of that moment, the bicycle traffic is a subsystem of the transportation network, and so on through the grouping levels.

The systems approach is generally applied to large problems. The common characteristic of the problems is that they cannot be viewed independently of their environment or contexts. For example, if the problem areas include education, welfare, law enforcement, and regional development, it is not possible to study these topics without considering the political and economic structures in which they are embedded. As a result it is frequently necessary to collect, organize, and use large amounts of data relating to both the phenomenon being investigated and its environment. Extensive data processing suggests the use of a digital computer, and this is one of several ways that the systems approach becomes involved with the digital computer.

As this opening discussion has moved back and forth among the systems approach, systems, and complex problems, the missing ingredient has been the outcome of it all. What about a "solution"? The solution to an application of the systems approach takes a variety of forms. Examples of solution forms include a plan for investing, a set of recommendations to a decision maker, better ways to use existing health care facilities, the design of a new organization to satisfy specific requirements, a modified highway network, and the design of a new piece of equipment to interface between a human operator and a task. The diversity of problems and solution forms is essentially unlimited. However, regardless of the nature of the phenomenon being investigated, the primary goal is always *increased understanding and insight.* If we assume that understanding of the phenomenon is achieved as a result of the application of the systems approach, the logical possibilities for use of this new insight are to evaluate existing conditions, to change and presumably improve the current status of the phenomenon, and to design and add something new to improve the state of the phenomenon. In all cases the outcome of the described process will be the solution, and the quality of the solution will be directly related to the quality of the application of the methodology. If the solution is a superior one, it will be defensibly so. For if the first attribute of the use of the systems approach is the achievement of understanding, the second attribute is the *visibility* of the solution attainment.

It makes a great deal of common sense to deal with a complex unmanageable problem by pulling it apart into manageable pieces. After studying the pieces and their interactions separately, reassemble them with suitable adjustments. In this way complex poorly understood phenomena are moved within the reach of human intellect and skills. For these appealing reasons the systems approach is sometimes referred to as "automated common sense."

Major Characteristics
of the Systems Approach (*1*)

The general structure of system methodology displays five major characteristics. Examination of these characteristics will move us more closely to the details of the systems approach. The systems approach is organized, creative, empirical, theoretical, and pragmatic.

Organized: There is scarcely an existing social problem which can be solved by the skills represented in a single profession or academic discipline. The issues of urban development that interest us are extensive problems whose solutions are likely to require the organized use of considerable resources. A wide spectrum of people possessing a variety of backgrounds, skills, and objectives is frequently necessary to study the problem area. The activities of generalists and specialists are needed over an adequate period of time to understand what is happening and what might be made to happen. The systems approach provides a kind of "glue" for the collaboration to be pulled together and aimed at the collective task. Implicit in the systems methodology is a common language, so that professionals in different areas can communicate about the same topic. The systematic nature of the various techniques facilitates a set of procedures that will allow each member of the team to function efficiently and to satisfy their own expectations. These last points, although made abstractly and somewhat low key, are intended to contain an optimistic note based on our experience in collaborations.

In any socially oriented research project, the value structure and belief system of the research-ers must be a significant dimension of the work. Often this consideration is difficult to accomplish with a small team, and with a larger group it is literally hopeless without assistance. All professions contain some historical commitments to certain styles in their work, and these may run counter to the needs of the team-solved problem. For example, the engineer has a penchant for quantification; the sociologist wants to form and test hypotheses; the epidemiologist collects extensive data—and somehow these kinds of focus must be directed to a common purpose. It has been our experience that the stability of such complex collaboration is most uncertain. It appears that little is known about the "care and feeding" of a long-term task-oriented team such as we have been discussing. The language, procedures, and style of the systems approach offer great promise for the successful operation of team problem solving needed for the solution of difficult urban problems.

Creative: Later in this chapter a set of limitations and cautions to be observed regarding an application of the systems approach will be discussed. In Chapter 2 we will review the extensive array of tools available to be used in an application. The selection of the appropriate techniques for a particular problem-solving procedure subject to a variety of restrictions requires creative effort on the part of the investigators. The systems approach is not a long series of routine steps which could be performed equally well by almost anyone. The necessary data base and theoretical support are seldom available, so that solutions must be found in the face of uncertainty. Furthermore, the solutions are not likely to be implicit in the data or the stated objectives of the system, so that only with great sensitivity can the solution be consistent with the ambiguous available information. It is very difficult to visualize future conditions in which proposed solutions must occur. Judgment must be exercised in the development and evaluation of alternative solutions. The problem area being studied must be very sensitively subdivided for conceptual manageability, and it is quite possible that the subdivisions will not be established along familiar professional, academic, or any other well-known organizational lines. In this regard the investigators may have to be courageous in their support of the innovative option. Not only is creativity called for throughout these steps, but the creativity must be supplied by a variety of talents and skills for which the systems approach provides a conceptual structure through which these creative energies may be focused.

Empirical: Much of the visibility of the methodology of the systems approach is supplied by great emphasis on data. The evidence for each conclusion must be as unimpeachable as possible. This stress on quantification is a mixed blessing, and the limitations will be presented later. Means for dealing with large amounts of data are available, and the problem-solving team will usually require skills for organizing and manipulating extensive data. The problem subject will be embedded in society and will interact with a variety of other organizations. If we are designing a rapid-transit system for an urban area, other organizations, such as municipal governments, labor unions, and local business groups, will exert an influence on our design. Their attitudes, beliefs, values, and practices must be considered. Although these effects are not easily quantifiable, we can try to utilize

procedures available to account for the influences. The nature of the analytical strategies employed in systems methodology provides guidance for the kinds of data we will need, and how it must be organized.

Theoretical: The two critical ingredients required for the operation of the systems approach are a suitable data base and theoretical support for the subject area being examined. All basic facts and theories must be collected. Invariably there is both a data and a theory gap. Limited theory can be dealt with only by imagination and intuition creatively employed. And, of course, a frequent by-product will be some additions to the existing theory. The organization of the relevant theoretical work for an urban phenomenon is likely to include material from economics, psychology, planning, architecture, engineering, sociology, hydrology, ecology, political science, and anthropology. Regardless of the strengths represented on the problem-solving team, access will be necessary to other people and sources with additional relevant capabilities.

Pragmatic: The intention of the systems approach is to have an impact on "real-world" affairs. Contributions to the understanding of social processes and to the relief of human problems are the primary goals, and any other results, such as theory development, are secondary. Whatever is done in the methodology is part of an attempt to respond to a real need.

Basic Steps of the Systems Approach

In order to move closer to the specifics of the systems approach, in this section we will classify the methodology into four basic steps. This classification is completely general and independent of the nature of the investigation. The range of activities the steps could accommodate include achieving understanding of the phenomenon, evaluation, deleting or changing components in an existing system, and designing new components and systems. A useful breakdown of systems methodology is: problem formulation, modeling, analysis and optimization, and implementation. These are not independent steps, and at any point the status of all phases must be in constant review.

Problem Formulation

Problem formulation requires a deep understanding of the total problem in the form of verbal descriptions which permit quantification of the significant features. Since the desired quantified relationships are to be used in the modeling and analysis that follow, their formation must consider the tools available for analysis. This first step is the most difficult, and it often requires as much as three-fourths of the total effort. Consider, for example, an investigation into some observed and perceived difficulties in the welfare processes of an urban area. If systems theory is to be used to help solve problems, the corrective actions to be applied imply that there must be decision makers. All decision makers must be identified, and their range of alternative actions must be determined. A truly complete list of alternatives is incredibly difficult to form. However, the successful compilation of a list of alternative actions simplifies the analysis that follows. After identifying the alternative actions, the consequences of each action must be determined in terms of the goals and value structure of the

decision maker in the welfare system. This is another very difficult task, and it must be accomplished with a superimposed awareness of just what can be quantified. And finally, having identified the alternatives and outcomes, the influence of the total environment within which the decision process occurs must be included, especially the economic and political dimensions of the problem.

Next, in the *modeling step,* one goes from the real world of the problem to the abstract world of the modeler. Forming the model is not a rigorous step which is done systematically. It is an artistic and creative step, and for a given phenomenon there are likely to be as many abstract representations of the real world as there are modelers. Whether one model is better than another can only be determined in terms of the total systems solution and how well it relates to real data collected at the site of the social phenomenon. The modeling step is a critical ingredient in the systems approach. Modeling requires a sensitive balance between including the most relevant aspects of reality in the model, and keeping the complexity of the model limited by the practical considerations of existing theoretical tools, computation time, and data availability.

The third step of *analysis and optimization* takes place in the model world. A great deal of systems literature is devoted to this phase, where the model is analyzed to find the best strategy for resolving the research problem within the domain of the model. The two options used most are analytical techniques and computer simulation. These will be discussed in more detail in Chapter 2. Computer simulation often provides a fast answer to a specific situation, but if a deeper insight into the workings of social processes is desired, simulation is less satisfactory. In the analytical approach, more general results are possible because explicit descriptions of the important interrelationships between the optimum strategy and the value structure and environment of the decision maker are calculated.

Implementation, the last step, is the procedure by which the results determined from the model are translated as a set of actions to the real world. Once again, it must be emphasized that actions taken in any step must be assessed in terms of their influence on the other steps. For example, public knowledge of the existence of the results of this last step may change factors contributing to formulation in the first step. A case in point would be a decision to rezone an urban area effective on a future date. The knowledge of this new information will influence the problem formulation and will have to be considered. In any case, the status of the social process being changed will be the indicator, either to validate the action taken or to suggest that a new action be taken.

An interesting alternative form of the four steps just discussed is based on a series of five questions. The answers to these questions suggest rather more specific tasks to be accomplished or actions to be taken. As the investigation into the phenomenon or problem begins we ask: What is the *state* of things? This is a nonevaluative question. It requires that we select the entities of the system and list their attributes and any activities in the system that can cause change. Then we must gather data that provide the values the attributes can have and define the relationship involved in the activities.

What is the *status* of things? The answer to this question requires evaluation. Performance must be measured against a visible accepted standard. In a simple example, if we are examining a business organization, the data from the first question might be in the form of a profit-and-loss balance sheet. The second question would be answered by whether the company made a profit or suffered a loss.

What is wrong (or right)? We must now do some form of analysis to determine where the difficulty is. Or if the evaluation were positive, it would be helpful to know the factors contributing to success.

How can we fix it? The existence of this question and its answer assumes that the evaluation was negative. If it were positive, the investigation is likely to terminate with the third question. To resolve the troubles in the system we must make a series of recommendations, cast in a useful form for the person or agency that has the power to act upon them.

How can we promote our results? In this last optional question, the investigators strive to achieve the implementation of their recommendations. This potential injection of the people doing the study into the policy-making process will be discussed in more detail later. The intent, however, is that the collaboration of source people doing the work will make a commitment for positive utilization of their results. In the social sciences this is often termed "action research."

Purposes of Modeling

In the vernacular of systems language, the process we have been describing is often referred to simply as *modeling*. It is now appropriate to be more specific in the answer to the question: Why model? Four answers to this question are listed next (*2*).

1. *The modeling process can be used to achieve insight and understanding in regard to the phenomenon in question*. The characteristics of the model can be manipulated to deduce regarding the real system many consequences one may not have been aware of. The abstract model permits ease in manipulating properties and definitions instead of the complexity of changing actual situations and entities. Starting with a primitive model based on preliminary observations and measurements, the model testing is likely to suggest new real-world experiments. The results of the new field or laboratory studies permit refinement of the model, and so on back and forth between the model and the real situation in a potentially progressive exchange not possible by each process alone.

2. *The model may be used for prediction*. After the model is able to produce historically verifiable results, it can be perturbed in a new way and the effect observed. In this way, decision makers can have some idea of the consequences of alternative decisions before actually making them.

3. *The model may be used to design organizations*. The principles for the design of organizations possessing certain properties and modes of operation can be established in advance by using arrays of models and systems analysis to examine alternative structures.

4. *Models can make contributions to the techniques of measurements in social research*. We will see in examples to be presented that the parameters of the model may be far more useful than those traditionally measured. An alternative version of a model may give insight into just what the investigators should be measuring.

Brief History

Any historical review of systems analysis must necessarily be brief, because the most obvious antecedent events occurred during World War II. However, Churchman in his recent book (*3*) suggests that Plato's *Republic* is actually a famous systems-science book. Plato thought that he could begin to design the underlying model of a city-state. From an examination of the political structure, inferences could be made about the justice system. Using the general notion that the objective and logical approach of the scientist would be an attractive and useful adjunct in decision making, Churchman points out that writings by such authors as Nietzsche, St. Thomas Aquinas, Descartes, and Spinoza among others may be considered as early systems studies. The more obvious direct historical tie comes from the activities in England of teams of scientists who were asked to research some of the difficult military operations of World War II. This "research into operations" eventually evolved into the discipline known in this country as *operations research*. The World War II operational research concentrated on solving tactical and strategic problems by optimizing the allocation of resources. These research teams of scientists were very successful, and following the war there was an attempt to apply the methodology to a host of problems both industry and the government wanted solved. *Systems analysis* as it evolved differed in scope from operations research. A much broader range of problems and a longer time perspective became important. Post–World War II problems that were considered included industrial production, marketing, finance, and transportation. In the early 1950s the availability of large digital computers with enormous data-processing capabilities gave great impetus to the data-oriented systems approach. The marriage of the computer and systems methodology marked the beginning of this contemporary period of problem solving. One of the best known consequences of this union uses the SAGE system. This system could supply, in a very precise and intelligible fashion, information about all aircraft and unknown objects in the air.

More and more scientific people were influencing policy decisions through the activities of a number of nonprofit corporations, such as Rand and Aerospace. These organizations were formed to study important strategic and tactical problems of the military. During the 1960s, when national attitudes toward military matters and the space program appeared to be changing, a substantial effort was made to utilize the systems approach in a broader range of problems. New client-sponsors began to appear.

An interesting early example began on November 14, 1964, when Edmund G. Brown, then Governor of California, announced the state's intention to call for bids from the aerospace industry to develop plans for four problem areas (4). The following quote by Governor Brown is given to show the nature of the statements used to elicit proposals from the proponents of systems methodology.

First, transportation. We will ask the systems engineers to study ways to provide a complete trans-portation network within the state, efficiently coupled into land, sea, and air transportation from out of state. We will ask them to identify the major patterns of movement of people, merchandise, materials, and food within the state. We will ask them to describe the transportation system which the state will need 30 to 50 years from now to provide efficient movement. And, finally, we will ask them to tell us how much such a transportation system will cost; who should pay for it; who should run it.

Second, we will ask the systems engineers to design new ways to cope with California's criminally and mentally ill. This is a problem with which it is becoming increasingly difficult for California to cope. Our population is growing and so is the population of mentally ill. There are flaws in any system that involves institutional control and we will ask the aerospace teams to suggest ways in which they might be corrected. Perhaps an entirely new social structure within a hospital is desirable. We would like to know whether the cost of care can be cut and the efficiency of treatment be improved.

The third problem we will pose to the systems engineers is that of accurate collection of information in which government and industry can base decisions for years and even decades ahead. We will ask the aerospace engineers to design systems that will improve our data on diseases and educational requirements. We will ask them to provide information on special needs of some of our population we might now be overlooking.

Finally, waste management. There is a system at present for managing the wastes discharged into the air, soil, and water of California as a result of consumption by men and machines of materials which are necessary to support life or to produce goods. But it is not a system which has been developed by deliberate design to meet the state's needs.

This was a substantial request of the space- and hardware-oriented engineers, and after discussing, in the next section, some of the problems associated with applying the systems approach, we shall return to consider some of the consequences of Governor Brown's charge.

Since the early 1960s many people, in all levels of business and government, have been involved in problem solving using systems methodology. The spectrum of professionals participating has included planners, operations researchers, engineers, management scientists, and dyed-in-the-wool systems analysts. In later chapters we will be reviewing some of the activities and results in the area of urban development.

Objections to the Systems Approach

Anyone utilizing systems analysis to attempt to achieve understanding or to solve problems associated with social processes is likely to find it necessary to cope with an ancillary set of difficulties. In addition to the intrinsic difficulty of the basic effort, a whole series of issues must be resolved between the investigator and his peers, the affected population, the client-sponsor, and the population at large. An entire hierarchy of problems, many quite unexpected, has been experienced by researchers in the proposal and application of the systems approach to problems related to social phenomena. Therefore, in order to realize the advantages previously cited, we must deal with the five categories of objections given next.

1. *Social processes are not amenable to the methodology of the systems approach.* Some believe that the techniques utilized in hardware systems cannot be applied to living systems because of an essential difference in the subject matter. This objection asserts the existence of a fundamental dichotomy between the natural and social sciences, but there is no concrete evidence that this is so. Part of the objection relates to the penchant of systems people toward quantification of variables. The opposing position is that social phenomena are simply not amenable to extensive and "hard" quantification. Some of the concern may be related to the difficulty of changing research methods and thought habits, and to a limited understanding of the new methodology.

Also related to this concern is the fact that there is no solution to social problems in the usual sense. One does not solve, for example, the problems of welfare. At an arbitrary point in time, one can work on aspects of welfare systems, but when one stops working, the system goes on and presumably into another state and a new class of problems. In addition, there is some indication that when the pressures of problems in one social system are relieved, the pressures are translated in other forms to other social systems. In this context, there is no solution, and the notion of right, wrong, true, or false solutions must be viewed in a different way from other types of problem solving.

2. *Social processes research may be used for the suppression of human freedom and the dehumanization of social life.* Social scientists have always had to be especially aware of all the consequences of their research. In a collaborative effort, all members of the group must also shoulder these responsibilities. If it is assumed that there are "laws of human behavior," and if an effort is made to determine them, some people fear that this is but an ill-concealed attempt to seek to control humans and thereby limit their freedom. This is an "anti-knowledge" position;

those who take it fear that new knowledge puts power in the hands of the discoverers and take a pessimistic position regarding the utilization of the new information. Our research must counteract any potential role in the dehumanization of society by taking an active role to ensure that it is humanization, not dehumanization, that occurs. If not, investigations into social processes will be paralyzed.

An aesthetic objection fears the diminution of the image of the human being as we understand more about him. For some people there is an attraction in uncertainty and mystery. A similar fear has existed for many people because of the accumulating information relating to the natural world. However, the increasing store of knowledge appears not to have diluted appreciation of the natural world; rather it has changed the basis of the admiration from a somewhat simplistic view to a far more significant level. Hopefully, understanding of human beings and their behavior will have this effect. Social research should not deprive man of his wholeness and unique individuality. The goal of the investigator, unlike the poet, is not to capture the richness of an individual's existence but to develop general propositions about him. The use of the systems approach presupposes that this knowledge will contribute to the advancement of human welfare, the rationality of social decisions, and the achievement of constructive social change.

3. *Social phenomena are too complex to be analyzed.* A very frequent objection arises from the complexity of social processes. The implicit assumption is that oversimplification in any social problem solving is unavoidable. Many research-

ers would agree that the most pertinent limiting factor is not the modeling methodology but the ability of the human mind to grasp a complex situation. The systems approach offers the possibility of breaking up a complex problem into an array of simpler ones, thereby placing the issue within reach of the individual's analytical powers.

Another difficulty, somewhat related to difficulties already stated, is the feared incompatibility between the rational tools of systems analysis and the frequent examples of the irrational nature of man. An example is the striking disparity between the powers of the intellect when applied to attempts to mastering the environment, and man's awareness of the consequences of these actions. Another example is the seeming impotence of man in dealing with many aspects of the conduct of human affairs. However, whatever the nature of human behavior, these features must be included in the analysis. Value structures and belief systems are extremely difficult to include.

Social systems defy definition as to objective, philosophy, and scope. For example, how can a welfare system be defined without including the systems of health, education, employment, and individual behavior? And all these interlocking processes must include the value structure and belief system of the diverse people involved, such as the administration, the recipient, the Black Power advocate, the social critic, and the politician. This objection is a graphic illustration of the complexity problem.

Neither the researcher nor the process of social systems research can be value-free. However, to retain scientific objectivity, a great effort must be made to account for the influence that the values, attitudes, and expectations impose on the research apart from the actual problem itself. The relationship of the investigator to the people and the social structures he is studying must be included as a salient dimension of the research. In addition, an unusual relationship may develop between the study team and their client-sponsor. Often the study of an urban issue is funded by an agency of the current administration, and the evaluation of the issues and recommendations for change may be very threatening to the status quo. Problem-solving processes that evolve into this mode are likely to be short-lived or to be discredited. The systems approach aspires to make visible the value structures and belief systems of the researchers, the affected population and processes, and the decision makers, and to cast its results in a form that can be interpreted by all three.

4. *Systems analysis has invaded many traditional social science areas in recent years with few noteworthy results*. Some social scientists display disillusionment and resentment because of the results of some contemporary systems analysis. Studies of crime, waste management, mass transportation, information storage and retrieval, and social welfare have been made without including appropriate disciplines, without adequately considering political and economic factors, and without forming the results in a manner suitable for implementation. As in any complicated technique, the skill of the people applying it has varied, especially in the early stages of the evolution of the technique.

5. *Involvement in research on social phenom-ena requires the participants to understand and support the ethical positions of the social scientist*. The group studying social processes faces a whole array of issues, some of which may not be familiar. When human subjects are involved, a number of ethical questions must be considered. Questions of deception, invasion of privacy, and misuse of findings become important. Previously discussed, but also pertinent here is the requirement of a constant awareness of the consequences of producing knowledge about the control of human behavior. The researcher must establish and assess his position regarding the potential contributions of social research to social action and social change. Other disciplines, along with the social scientist, may face the ambiguities inherent in their involvement in the policy process. And a new view of scientific objectivity must be developed for the study of man and society.

These five objections just cited represent a complex, bewildering, and almost overpowering plethora of problems associated with the application of the systems approach. We have presented them to emphasize that the individuals in the study teams must consider these issues and find a means of forming attitudes and philosophies which will permit them to function in the social problems area. Consider the issues and then, hopefully, get to work. It is clear that man is confronted with human and social problems of such depth and scope as to make one despair of finding solutions. In spite of the flood of literature each year devoted to analyzing the problems of the cities, race relations, poverty, unemployment, alcohol and drug use, suicide, delinquency, and education, the gap between this work and practical solutions to the problems is still wide and unbridged. We believe the systems approach shows promise for beginning to deal with these problems.

Cautions

After being alternately positive and negative regarding the utility of the systems approach, we now want to present a number of caveats. These cautions will attempt to discredit a number of "folk myths" that have emerged regarding systems methodology. We will demonstrate that although the potential of the systems approach is substantial, it has limits. The following list is given in that spirit.

1. Systems methodology will not give a simple set of procedures to arrive at incontestible conclusions. In Chapter 2 we will examine many of the very extensive set of tools and techniques available. In each problem, the use of these tools is likely to be quite different. As emphasized before, the strategy for selecting and applying the tools to a particular problem is a very creative process. The openness and visibility of the methodology will make clear the basis of each conclusion but will not guarantee the lack of controversy regarding the details of achieving the result.

2. There is no assurance that a given task can be accomplished within the given budget and manpower; however, for the incomplete part of a study, planning will be done so that priorities for further work will be established.

3. A common belief is that systems methodology serves as a kind of an umbrella under which a technique from another discipline can be transformed to be suitable for a new application. NASA has made substantial efforts to show that many of the processes, devices, and techniques developed for the space program have relevance in the civil sector. It is important to realize that if the modification can be done, it may take considerable effort.

4. We cannot be sure that our analysis will give a complete or totally integrated model of reality. The nature of the phenomenon may be beyond the capability of the modeling tools available. Until a methodological breakthrough is made, we may have to settle for a compromise combination of existing modeling structures.

5. A fear or fantasy, depending on the position of the observer, is that the systems approach will replace judgment. Instead, as we have said before, its goal is a better-informed decision maker.

6. Similar to 5, some people believe that systems methodology can replace the political decision process—that the rational, logical approach should be a necessary appendage to a democratic process. Again, however, the intent is to provide better information with a visible basis.

7. Client-sponsors of systems studies cannot have a passive role during a study. There must be active and continuing participation of the client organization.

8. The systems approach does not provide a value system that implicitly implies objectives to be satisfied. Although the methodology is rational, the social process need not be. If the values of the investigators become involved, an evaluation must be made. All evaluations, standards, and performance measures must be very explicit and on display.

Variations in the Systems Approach

Of major importance in this introductory chapter are the fundamental ideas of the systems approach, but we are attempting also to get the reader to think in a systems way. And it is here, especially, that our bias is showing. We have been describing just one of several variations in the systems approach. It is the one most in use currently, and the focus of the version is the use of the "scientific approach." The scientific version has generated many interesting ideas and techniques recently and is, perhaps, the best known. Beginning with the science version of the systems approach, the scientific flavor is best characterized by the use of a mathematical model. The mathematical model has a particular set of attributes of the real world cast into a form consisting of four parts:

1. The concepts or variables of the model, including both primitive and defined terms.
2. The postulates, which relate these variables in some fashion.
3. The mathematical operations performed on the postulates to obtain deductions.
4. The deductions or theorems, which derive from the postulates.

The mathematical model is intended to be *objective,* and other versions of a model in the scientific systems approach are economic or behavioral. The emphasis, in any case is on logic, rigor, and precision to the extent that is possible.

The second type of systems approach might be termed "efficiency." The emphasis is on spotting difficulties in a system, including poor performance, failure, and waste, and then attacking these sore spots to remove them. The premise is that when a process is working well, it is difficult to learn much about it. However, an understanding of the basis for any marginal or critical operation leads to an understanding of the total process. One extreme example of this viewpoint is a methodology that studies only the accidents and failures of a system. Data on catastrophic episodes are used to achieve general understanding and to formulate recommendations for improvement.

A more typical efficiency approach, however, takes the position that there is a best way to do a particular thing and that all efforts should go toward approximating this optimal behavior. The obvious difficulty, and the basis of most criticism of the efficiency approach, is that the one best way of operating subsystems may be inefficient from the viewpoint of the overall system. History is filled with examples of this. A complex organization may have the need for the accomplishment of many repetitive tasks, and the people performing the tasks could be subjected to a time-and-motion study and replaced by automatic equipment. The general effect on the personnel of the organization may be more detrimental than the gain in efficiency of the individual operations.

Perhaps the greatest difficulty in implementing the efficiency approach is the absence of overall system performance measures. The relationship between total system and subsystem performance is seldom so clear that the manager can resist across-the-board cost reductions or other attempts at waste limitations. The proponent of efficiency generally believes a total systems viewpoint to be too idealistic. When a source of trouble is determined, the difficulty should be resolved. The user of the scientific methodology who searches for general propositions would agree as long as it is clear in terms of the total system what is really inefficiency or trouble, and what is not.

The efficiency supporter's counterargument, pointing out the complexity of the science position, is not without merit. It is possible to win the philosophical debate and fail to solve the practical problem. Although the efficiency solution is limited and offers little general learning, it may have to be utilized in particular problem areas where the scientific approach appears inappropriate.

A third version of the systems approach emphasizes the human aspects of a system. The thinking of the scientific approach is, in a sense, turned inside out. Instead of the nature of the total system being a significant determinant of individual behavior, the system is thought of as a compendium of the behaviors of individual people. In this way emphasis is shifted to human values: freedom, dignity, privacy. Questions of performance at any level are kept out of the systems characterization. The difficult issue, always prevalent in the social sciences, of how to collect data on human behavior without changing the behavior, arises at this point. Laboratory experiments, case histories, questionnaires, and interviews are all used to collect data, and all methods have their detractors who will make a case to show that the data are invalid.

The behavioral scientist wants to consider the nature of the individual person, social group, society at large, and cultures. For people and groups he studies beliefs, concepts of reality, and value structure. He tries to find the answer to the question: What is it that people want? The decades of study by the social scientist have been able to answer this question in various ways, and these answers provide a basis for examining systems.

In these brief reviews of alternative versions of systems methodology, the last to be considered is the antiplanning approach. The descriptive name is self-explanatory, and the philosophical variations of this viewpoint are many. In all cases the notion of using a scientist or behaviorist to analyze or change a system is rejected. Antiplanners may believe in the operation of the system by an intuitive manager who is experienced, bright, and educated within the system. And this man will be judged by his performance. The "self-made" successful men of popular history are examples; however, an in-depth analysis of the careers of many will give confusing results regarding their decision-making ability.

The skeptic is an antiplanner. He believes the nature of the real world is a mystery, and he discounts activities that lead to growth, progress, change, comfort, or convenience. Others believe that society is not markedly affected by human decision makers but shaped by continuous sociological mechanisms beyond our control. There is a long historical record of thinkers who believed that all events are predetermined.

In a related way, the religious approach takes the position that a supreme power is responsible for all real planning of the world. Hopefully (but not in all religions) this is a force working for good. Here, of course, human behavior will be adjusted to what the supreme plan is. Frequently in our culture the scientist has separated his beliefs as a man regarding the existence of God from his activities as a scientist. This, in many cases, justified ambivalence is bound to have an effect on the planning aspects of analysis. The main point of all this from the scientific-approach position is that if religion is a significant dimension in the process being examined, it must be considered.

Perhaps the most important point regarding the efficiency, humanist, and antiplanning aspects of the systems approach is the fact that any system created or modified by the scientific approach may be irrelevant or even damaging to those people in the affected social process who live by another philosophy. Somehow all attitudes must be determined, made visible, and accounted for in the analysis.

Example

This book will be dominated by the examination of cases where the systems approach was used to deal with issues in the urban setting. Most of the problems will be current, and the solutions are on-going. In this opening chapter, and prior to the presentation of analytical tools in Chapter 2, we have chosen to discuss, in a descriptive way, a completed systems study which was referred to earlier. This particular study, in our opinion, is a kind of a historical milestone in the development of systems methodology. The project was the development of a California state information system by an aerospace company at the request of former Governor Brown (3). The intent was to provide all decision makers with information of the highest quality possible, quickly, and in a usable form.

We will look at this project using the steps of the systems approach.

First, in problem formulation, the objective of the state information system must be specified. Three sentences ago we gave an imprecise layman's version of the objective. The system designers specified the objective as ensuring that the new computer-based system was better than the existing information system. The state of information gathering and data availability is near chaotic in many cities and states. And even when data are available, decision makers do not know how to deal with the mountain that soon accumulates. An improved system meant lower cost, more information in a shorter time, and any other improvements possible. The obvious critique we can make here is that the *standards* of the old method are to be maintained but implemented in a new way. Implicit in the concern to meet existing needs in a better way is the assumption that short- and long-range needs are not likely to change much, and that capacity for innovation at any time will be limited. Also assumed in the stated objective is the absence of possibility for much change in the procedures and policies of the several user state agencies. We are making these observations as examples of visible exposition which should accompany a foundation block such as the system objective.

Another part of problem formulation is the detailed description of the environment in which the information system will operate and the resources available for operation. Environment includes such factors as budgetary limits, actual information currently being transmitted between and among agencies and the public, potential for adding or deleting uses, and the possibility of including new types of information. The resources for the new system include personnel, money, skills, and information files which already exist. New resources include the modern design methodologies, data-processing systems, computer hardware, and the actual political leverage available to assist in the evolution of the new information system. A negative resource, and unfortunately one not considered in this study, was the political resistance that existed in the state during the study.

With the system analysts aware of the objective, environment, and resources, the next effort was the consideration of alternative models to represent competing designs.

Hundreds of interviews with local and state officials were held to determine the handling of such items as auto licenses, crime reports, job placements, tax returns, and other vital statistics. Estimates were that over 400 miles of filing cabinets would be necessary to store all the information by 1980. In plotting the flows of information, the investigators determined that there were over 1000 junction points forming an information grid in the state. Almost half of the state's departments and all the larger counties were using computers, but in a completely uncoordinated way. Paper exchange was still the major mode.

The more obvious hoped for savings in the elimination of paper work, duplicate records, and additional personnel were complemented by the possibility of contributing to the solution of social problems. With solid facts available, agencies could deal more effectively with problems of crime and delinquency, health and welfare, education and welfare. For example, job openings could be matched to job needs anywhere in the state in a short time. Prerequisites for urban planning, such as land-use statistics and economic and traffic trends, could be available on an up-to-date basis.

With the information at hand and detailed knowledge of the "state of the art," the modeling process is literally a constrained form of sustained "brainstorming." Proposed designs included both centralized and distributed computer information storage points tied to the users by a variety of data-processing linkages. The idea of actual centralized storage of information was eventually abandoned because of the bulk and variety of information and the large number of users. A hybrid model was conceived which located pools of data stored at various agencies with a centralized computer locator service which could be bypassed if the appropriate source pool of information were known.

With this basic structure agreed upon, a number of system components could be identified and a variety of specific tasks defined. The central information locator could be designed and fixed geographically. The information pools could be spatially located, and the types and methods of filing information specified. A variety of linkage means were necessary to tie the users pools to the locator. Users would operate from small consoles or terminals and query either a computer or a programmer via telephone lines or microwave links.

The quality of the performance of the proposed structure could be detected by the following:

1. The number of errors the central locator made.
2. The number of errors the information storage pool made.

3. The ease, accessibility, and accuracy of the user terminals and linkages in transmitting requests and returning answers.

These performance measures are, however, related to the system design itself and do not include consideration of how well such a system could be integrated into the activities of the state. Information has utility and value in several ways, and this new system would change many modes of transactions of this valuable commodity.

The proposed information system was not implemented because of a significant change in the environment of the state; however, let's look at some of the potential difficulties implicit in the design.

1. No means were provided for managing the new system. The emphasis was on the technical aspect.
2. Political resistance was not evaluated.
3. The designers did not concern themselves with the implementation of the information system.
4. No general evaluation criteria were given.
5. No serious study of information itself was done. Such a study would include:
   a. What information is really needed and how often?
   b. What criteria should be used to determine information accessibility?
   c. How do you handle information that invades citizens' privacy, such as health and criminal records?

The emphasis in the study, for a number of reasons, was on utilizing new technologies. It appeared that the administrators who contracted the study and who were to use the system played too passive a role. Even if implementation of the system had begun, it is likely it would have been discontinued since the next state administration took the efficiency approach and imposed substantial financial constraints. Should the study have predicted this substantial change in the environment? It would be difficult to chide the designers for not being prepared for all political eventualities. In any case, the intrinsic promise of the study was never realized. At this point in time the basic issue is still there. How do we provide our decision makers with the best possible information, quickly, and in the most usable form?

Public Policy

To conclude this introductory chapter we want to discuss the promise of systems methodology in terms of its contribution to an evolving and emerging new field of activity—social policies planning. Many forms of planning will be discussed in this book; however, one of the most exciting possibilities for contributing to societal development is appearing because of a fortuitous combination of circumstances and events. As social problems have increased and become seemingly insoluble, more and more individuals in diverse settings have sought multifaceted approaches to solutions and new developments. There is no real attempt to break down professional and disciplinary boundaries; however, there is growing recognition that no one social problem can be solved by the skills available in any one discipline.

The description of this new field of inquiry is cast in a form using systems vocabulary and ideas we have already presented. The three basic ingredients are a modeling strategy, program planning, and the concept of development (5).

The modeling strategy for social policies planning came from the efforts of the ecologists and other environmentalists to develop models that represented human and social systems. Such a complex system is composed of interacting and interdependent subsystems which are open to and adapt to their environment. The processes may be described as self-organizing and self-regulating. The model is clearly an alternative to the mechanistic model which dominated Western thought for over two centuries. The mechanistic model portrays "machine-like" behavior in which the internal state and the state of the surroundings is known precisely, and these define uniquely the next state to which the system will go. The ecosystem model came from a diverse mix of sources, including independent scientific developments, public programs, and popular movements. Specific findings such as data, models, and methods have converged to form an early generation model of considerable sophistication.

Complementing the ecological systems modeling strategy, the second feature of social policy planning is program planning. Program planning refers to an alternative mode of planning in which the process of adaption of ecological systems to their environment is emulated. Concepts and methods are available for dealing with system deficiencies and malfunctions while directing future development in desired directions. Program planning is also a hybrid with roots in economics, administration, engineering, planning, and management science. It is quite different from other types of planning, in which attempts are made to describe desired end states.

The concept of development may be viewed within the framework of a paradigm. The several services associated with urban systems are usually viewed as independent processes. Each service represents the bias of one profession, is oriented to a particular activity, and intervenes in on-going urban processes in a certain way. Clearly, this is true of medicine, education, city planning, public recreation, housing, and so on. In this same way, the taxonomy of federal agencies imitates that of the traditional professions. Appropriate professionals lead these agencies, such as those for mental health services, housing, and welfare, and the organizations are in competition with one another for budgets, prestige, power, employees, goals, and clients.

Development refers to the processes by which complex societal systems interact with their environment to accomplish growth, solve problems, achieve stability, and exploit opportunities by accumulating information, capital, and functional capability. The core of the social policies planning is taken as maximizing individuals' development while concurrently improving the total social system.

Following the complacent 1950s, chronic urban problems of the 1960s have made it patently obvious that the professional bureaucratic jurisdictions do not coincide with the structure of social systems or with the nature of either personal or social problems. The implication is clear— problem solving must be accomplished within a developmental framework; that is, there must be system understanding, causal relationships among subsystems, and governmental structure which allow the mounting of all necessary resources to attain precisely stated objectives. This suggests a change in strategy which allows instituting desired conditions, instead of attempting to deal with the symptoms of undesired conditions.

Most of us are observers, not participants, in work being done to develop this new view of public policy. However, the new planning is as conceptually attractive as the systems approach. As a parallel example, many thinkers in the field of medicine believe that the emphasis should be shifted from a buildup of skills and institutions for caring for the sick, to efforts to prevent illness before it occurs. Instead of focusing on ways to medically intervene and cure a sick person, the altered position would force recognition of the fact that poverty, in some areas, is a major causal factor for illness. Expanding this point, a county health officer would feel responsible for finding ways to improve employment opportunities, raise income levels, and increase the positive aspects of every dimension of the total environment which affects health. Here, good societal conditions are clearly implied, and a traditional boundary must be overcome in order for the health professional to function in this new way.

The basic notion of this public policy is to consider the integration of the public services rather than their segregation. In the areas of education and economics, the practitioners have learned to see themselves in a cooperative venture with the several systems with which they interact. For example, national fiscal and taxation policies reverberate in a series of actions and reactions influencing the behavior and growth of an incredible variety of organizations and individuals.

There are a number of contemporary intrinsic limitations in the actual implementation of social policies planning. Many of these limitations are related to the objections to the systems approach given in an earlier section. Methods in social and behavioral science are revealing unanticipated complexity, social change theory is inadequate for valuable prediction, and there is the ever-present limitation in data base and intellectual capacity.

Fried in reference (6) utilizes the following quotation to hit to the heart of his position:

The technocrats, systems analysts, and economists, whom Reich puts into Consciousness II, work on the assumption that the difficult and interesting task for intellectuals and planners is the task of maximizing in some way the satisfaction of the tastes and values in society. The content of these tastes and values is thought by them to be beyond serious intellectual consideration and analysis. To be sure the content of people's tastes and values might be a subject for sociological or psychological analysis, that is, for studies indicating the way they are formed and how they affect conduct. But a wide variety of modern intellectuals think it heresy that values should be studied as such, and ends which both the student and his reader hold or should hold. Galbraith and others have noticed and criticized this technocratic bias by attacking its assumption that tastes and values are exogenous to the systems these theorists describe and manipulate. Instead, such critics as Galbraith stress how values are indeed endogenous to the systems which are being studied and worked, that is, these systems themselves generate and modify tastes and values. The glee with which these critics point out the fallacy of such theories is understandable; it is indeed a deep fallacy. The attractiveness of the fallacy to so many educated and serious people trapped in it, is that it gives intellectuals useful but fundamentally unassailable tasks to perform. Intellectuals concerned with the social system become technicians: if their work is faulty, it is faulty for technical reasons. They are not responsible for the values and tastes they are seeking to maximize since these, the fallacy teaches, are given—external to the systems they devise and manipulate—and in any case beyond rational discourse.

This counterposition debates, to say the least, the validity of systems planning for social policy. The term "debate" is a good choice to describe the several counterpositions which are sprinkled throughout this chapter. Our intent has been to inject the criticism into the very context in which the methodology was being discussed. However, it is our position that the debate is largely an intellectual exercise, and regardless of what approach is used to social problem solving, or what version of the systems approach is implemented, the solution is likely to be in error, perhaps seriously. The real issue and ultimate value is in the *progressive aspect* of performing the cycle of examination, solution, and evaluation, and then repeating it after having learned something. This is true for a wide variety of methodologies, but evidence is accumulating and a thought-provoking case can be made for the systems approach. Each day in innumerable social settings decisions are being made at every level which influence the health and welfare of our citizens. We take the practical position that the systems approach is a good idea; and it is ready to be applied now to help make those decisions better ones according to some general agreement. Each professional who will participate in this process must confront the issues we have discussed, form attitudes and philosophies that will permit him to function—and then, hopefully, get to work.

References

(1) M. Adelson, The System Approach—A Perspective, *SDC Magazine,* **9**(10), (1966).
(2) D. J. Bartholomew, *Stochastic Models for Social Processes,* John Wiley & Sons, Inc., New York, 1967.
(3) C. W. Churchman, *The Systems Approach,* Dell Publishing Company, Inc., New York, 1968.
(4) I. R. Hoos, A Critical Review of Systems Analysis: The California Experience, Paper 89, Rec. 1968, Space Sciences Laboratory, University of California, Berkeley, Calif.
(5) M. Webber, Systems Planning for Social Policy, taken from *Readings in Community Organization Practice,* R. M. Kramer et al., eds., Prentice-Hall, Inc., Englewood Cliffs, N.J., 1969.
(6) P. Nobile, ed., *The Con III Controversy,* Pocket Books, Inc., New York, 1971.

CHAPTER 2

Tools of Systems Methodology

Intellectual Borrowings

The array of tools, skills, techniques, and methodologies that may or must be present in a problem-solving collaboration is seemingly endless. When we examine a list of them, as we soon will, it is apparent immediately that many individual tools represent large, well-established disciplines. In almost all cases, the tool is the result of a long historical effort, a substantial literature exists, and the subject is frequently treated in a book form. For all disciplines the list will represent what might be termed "found tools," that is, tools which represent techniques and ideas that arose in fields parallel to or related to the designated reference field. This occurs because of competitive success and analogous concerns, and often little effort is required to translate the tools into the vocabulary of another field and adapt them for use.

Frequently general concepts and good ideas arise and evolve in a field but in a way that causes them to be festooned with the jargon and vocabulary of the parent field. Because of this, the generality may be difficult to detect by other potential users. There is almost a sense of ownership associated with tools in some disciplines, and great resistance arises regarding the use of found tools. This is a manifestation of an anti-interdisciplinary position. The consequence of such a position is that every new problem has to be forced into a form that is amenable to solution by the "in-residence" tools. We may view this as a grown-up version of: "Give a small boy a hammer and he will find that everything he encounters will need hammering." An example would be an investigator who becomes convinced that every process has uncertainty associated with it, and therefore should be modeled from a probabilistic viewpoint. The uncertainties may not be dominant or, even if they are, they might better be handled in other ways for many classes of problems. Therefore, the subject of this chapter is intellectual borrowings. Our intention is to say something about many of the tools of systems methodology. This will provide, at least, an *identification* of the tools. To become a user or to participate with other users of a specific tool will require substantially more investment of effort by the reader.

When we have identified the several tools of systems methodology, it will be apparent immediately that none deals holistically with people, urban structures, or the total environment. And again it will be obvious that some means is necessary to utilize these various tools in such a context that the total system may be understood.

Classifying Systems Tools

As a means of establishing some order among the myriad of systems tools, we will use a classification scheme that evolved from a survey of systems courses presented at universities throughout the country (*1*). While this list represents the systems engineering side of the methodology, the interdisciplinary nature of the subject includes everything and will serve to establish our *vocabulary*.

Interdisciplinary Theory

General system theory
System analysis

Theory of System Structure

Topology of systems
Graph theory

Foundations of systems science

Linear systems
Nonlinear systems
Dynamic systems
Distributed systems

Mathematical Foundations

Modern algebra
Linear algebra and matrices
Topology
Complex variables
Integral transforms
Vector calculus
Functional Analysis
Differential Equations
Mathematical logic

Control Theory

Feedback-control-system theory
Stability theory
Optimal control
Nonlinear control systems
Sampled-data control systems

Communication Theory

Information theory
Coding theory
Signal theory
Detection and estimation theory

Cybernetic Theory

Artificial intelligence
Pattern recognition
Adaptive and learning systems

Flow in nets
Sensitivity theory
Multilevel system theory
Network theory

Stochastic Theory

Probability and statistics
Stochastic processes
Reliability theory
Statistical decision theory

Optimization Theory

Dynamic programming
Linear and nonlinear programming
Direct methods

Simulation and Experimentation

Numerical analysis
Analog and hybrid simulation
Digital simulation
Modeling and identification
Design of experiments
Instrumentation
Systems laboratory

System Design Methodology

Problem definition
System evaluation
System integration
Design for reliability and maintainability
Computer-aided design
Large-scale system design
Systems management
Engineering economic analysis

Cybernetic machines
Synthetic behavior systems
Mathematical theory of the
  human operator
Man–machine systems

Interdisciplinary Technology

Systems engineering
System simulation and
  synthesis
Systems design
Simulation and optimization
  methods

Hardware Systems

Control and communication
  systems synthesis
Circuit analysis and
  synthesis
Utilities
Vehicular systems
Computer hardware
Energy-conversion systems

Biological Systems

Organic systems
Bioengineering models
Cognitive processes
Man–machine systems
Neural nets
Human factors
Biological control systems
Biosystems

Applications

Applications of engineering
  models
Applications of control and
  systems theory
Systems analysis
  applications

Ecological Systems

Environmental systems
Water resource systems
Environmental bio-
  technology
Urban environmental
  engineering

*Computer Information
Systems*

Programming languages
Systems programming
Logic design of computers
Automata and switching
  theory
Real-time systems
Information systems
Discrete systems

Operations

Operations research
Industrial engineering
Transportation systems
Inventory-control systems
Quality control
Queue theory
Industrial dynamics

Socioeconomic Systems

Economic theory
Game theory
Utility theory
Urban systems analysis
Decision and value theory
Forecasting

The methodological resources of this list offer impressive potential for complex problem solving. Our task now is to determine what some of the most appropriate tools are and to learn something about how they are used.

Systems Classification (2)

Many of the systems methodologies have evolved because they were developed for a particular type of system for which analysis was required. Therefore, a review of the various types of systems will expose many basic tools.

A fundamental classification scheme for systems is usually based on mathematical properties of the equations that describe the activities of the system. The most basic is the *linear system,* which is described by linear differential equations. Such an equation is designated linear because all derivatives, or rates of change of the system inputs and outputs, are raised to the first power only, and there are no products of derivatives in the equations. A consequence of the satisfaction of the linearity condition is a system property called *superposition*. Superposition allows us to take a hard-to-handle input, break it up into its parts, and find the system response to each of the input parts applied separately. Then the total response to the complex input is found by combining the individual responses from the pieces of the input. This is a very reasonable simplification, but it is valid only when the system is linear.

Symbolically our linear system with input x and output y may be represented by Figure 2-1. The linearity property allows us to write the following statements:

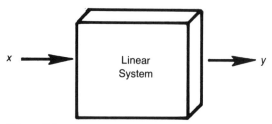

Figure 2-1

1. If y_1 is the response to x_1 and y_2 is the response to x_2, then $y_1 + y_2$ will be the response to $x_1 + x_2$.
2. If the model of a complex organization behaves linearly, then we can analyze each of its functions separately and combine the results to determine overall organizational performance. If we cannot do this, the system is *nonlinear,* and the necessary mathematical procedures to be used will be much more difficult, and may not exist. Most real-world phenomena are nonlinear, and linear approximations are frequently used to deal with them. A physical way to think of nonlinearities is that a linear system usually results if none of the components in the system changes its characteristics because of the *magnitude* of the inputs applied to it.

A system can be characterized as being either *fixed* or *time-varying*. Fixed systems are a consequence of the parameters of the system being unchanging with time. A long-distance telephone circuit changes from day to night as a consequence of temperature changes. Peoples' attitudes and organization procedures change because of the alteration of a multitude of external conditions. For time-varying systems such as these, the mathematics is again very difficult.

The parameters used to measure various aspects of the components of a system are classified as *lumped* or *distributed*. The issue is: Is there any change in a measurement due to a spatial variation, or can the parameter be considered to be lumped and concentrated at a point? Most complex systems are a combination. For example, in the closed stacks portion of a library, the checkout desk may be thought of as a point source of books, whereas in the browsing stacks, the books come from a distributed source. The cross-country transmission lines of an electric company are an example of a distributed parameter system. Whenever it is possible and the approximation is acceptable, the mathematical manipulations are much easier in the lumped-parameter case.

Another system description depends on whether the processes of the system are operating *continuously* in time or if they occur only at specific *discrete* moments of time. The mayor of a city provides full-time service while city councilmen usually meet only at regular time intervals for fixed periods. The best-known hardware system that operates only in discrete time is the digital computer. The mathematics associated with a continuous time system are generally more manageable.

Two more systems vocabulary terms are *instantaneous* and *dynamic*. If the system's functioning depends on anything that has happened previously, the system is said to be dynamic. If there is no memory, and the response at some time depends only on the inputs at that time, the process is instantaneous. Many simple repetitive tasks, such as filing and typing, can be performed in the instantaneous mode. If a home has a heating plant that is on only when turned on, and off only when turend off, the system is instantaneous. However, the addition of a thermostat and its operation based on a predetermined temperature setting changes the heating system to a dynamic mode.

Frequently parts of a system will not be known with certainty. The traffic flow at a particular point and time, the votes cast in an election, the weather on a Sunday are all examples of things we cannot know with certainty. The presence of uncertainties classes a system as *nondeterministic* and requires the use of probability and statistics to describe the processes. A system where variations are known with certainty is called *deterministic*.

As a natural part of the problem-definition phase of the systems approach, a complex system may be classified according to the taxonomy just discussed. This is a useful systematic way to start to examine the phenomenon, and, in addition, the classification implies what mathematics tools will be necessary, and possibly even what form a primitive model might take. A review of the list of topics presented earlier, especially under the categories of interdisciplinary theory, mathematical foundations, and stochastic theory, will show the subjects related to the vocabulary just discussed.

Control Systems (3)

Another characteristic of many systems is that, by means of various mechanisms, the output is monitored, fed back to the input, and, based on this information, the input is varied to control the output. Our eyes perform this service as we drive a car. The feedback control system is based, then, on comparison of output to input in an appropriate way, and control is achieved when a correction action is begun because the comparison was not suitable. Feedback may be continuous or discrete, and increase the error or decrease it.

Because this feedback and control process is present in so many phenomena, we can utilize a hierarchy of control levels as a further means of system classification. If we can identify the system control level in addition to other classifying statements, then selection of diagnostic tools and procedures can be done, the need for simplification measures is pointed out, questions are suggested that should be asked in determining system performance, and the location of trouble spots is facilitated.

The first level, designated *level zero,* has no control mechanism and corresponds to the instantaneous system previously discussed. There is no feedback and no memory. The home heating plant operated by an on–off switch is a level zero system. The simplest feedback arrangement is given symbolically in Figure 2-2. In our home-heating example, the process to be controlled is the furnace converting fuel into heat. The intention of the *level one* control system is to compare its output against a desired input and correct for *immediately* observed differences because no memory is involved. This is accomplished by the addition of a thermostat which turns the furnace on and off at temperatures above and below the desired setting. The consequence is an average temperature near the set value. Note that the furnace does not operate continuously; therefore, this level one system is discrete. We could have a level one continuous system by varying the fuel flow to the furnace to control the temperature. In summary, the level one system has no memory but provides a basic "tracking" ability with no way to make predictions or to make a conditional choice of actions.

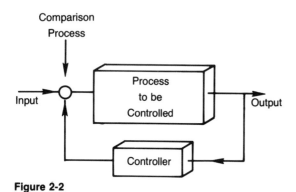

Figure 2-2

We can go to a *level two* system by providing memory. If we would like a lower temperature setting at night than during the day, and we store these settings by some means and allow a clock to select them, then our home-heating system is level two. The memory function must provide the capability to determine the necessary course of action based on the input information. Thus the level two can handle a wider variety of input conditions and can select among predetermined decisions about them.

To be able to identify a system as *level three* requires a system which performs all the functions of the previous levels, but, in addition, has the ability to learn. For a system to learn means that new plans can be developed, new decisions made, and adaption can occur to accommodate new conditions. Basically the system functions as a lower-level system and then, through a complex feedback mechanism which does corrective selection and plan development, the level three is established. The system requires a larger memory, a more complex structure of operations, and a more flexible memory organization. An example of a level three system is best given as one part of the composite of levels which nearly all large organizations possess. In a large manufacturing firm, the shop workmen performing the basic tasks of manufacturing and the salesmen who take orders for the product function within an organization as level zero. The level of supervision above the workmen and the salesmen is usually a level one system. The network of managers who control the shop foremen and the supervisors of salesmen constitute a level two system. With these preliminaries, we can now designate the company president and his staff as a level three system.

It follows, then, that the highest and most complex level is represented by the company's board of directors, and they are *level four,* a goal-changing system. Some of the functions that can be accomplished by level four are: develops new problems to solve, learns and then adjusts goals, corrects methods of learning, innovates, and actually controls the process of selecting and changing goals. Within level four all the other levels operate and are controlled. A very large and complex memory and extensive data needs are necessary.

The level classification used in combination with the systems vocabulary of the previous section provides a useful identification procedure for systems, especially with regard to accomplishing the modeling process. A separate discipline exists literally for many members of the several types of systems we have reviewed. System identification implies that we know something about the skills that must be present in our problem-solving collaboration, data and other kinds of support such as a computer, and the literature that is likely to provide the most methodological support. Much more could be said about other system classification methods; however, our intention in this chapter is to work from the long list of methodologies given earlier and show various ways that subgroups of the list can aid us in complex problem solving.

Linear Programming

Continuing to explore topics from the methodology list, in this first of several short sections, we will examine linear programming from the optimization theory category. In these several fairly brief sections, the intent will be to learn what the purposes of some systems methodologies are and how the methods are used.

Linear programming refers to one of a set of computer-based mathematical programming techniques. Many of the techniques are available as packaged computer routines, so it is not necessary to understand them in detail. Rather we would like to appreciate their characteristics in order that they can be useful in problem solving.

The general form of the mathematical programming problem is to find values of some quantities which combine to give us a result we would like to maximize or minimize. And this process must be accomplished subject to some specific constraints. A simple example would be to maximize the output of a multistep manufacturing process with a given budget. Mathematical programming uses a digital computer to repeat a search procedure until the best combination of factors is found subject to the limitations specified. The several mathematical programming techniques are differentiated according to the particular set of assumptions about the nature of the process to be optimized and its constraints. The matching of the programming assumptions to the specific problem allows a variety of problems to be analyzed.

The assumptions for linear programming allow the optimization of a wide range of practical resource-allocation problems. The linearity requirement restricts the form of the equations that relate the factors.

Let us consider a simple example to show the method of linear programming (4). A fleet of trucks is being formed to satisfy specific transportation needs between two sites of a corporation. The cargo needs are interpreted as requiring trucks with carrying capacities of 2 and 5 tons. At each site there is an array of loading docks and crews to handle the cargo. Incorporating this information into a problem statement we have $y = 5x_1 + 2x_2$, where y is the tons of cargo which we want to maximize, x_1 the number of trucks of 5 tons capacity, and x_2 the number of trucks of 2 tons capacity. The capacity of the loading docks and the number of men in the work crews at the two sites limit the number of tons of cargo that can be transformed. That constraint is represented by two equations:

$$3x_1 + 3x_2 \leq 18,$$

$$5x_1 + 3x_2 \leq 15,$$

where the symbol $\leq$ means equal to or less than.

We know that x_1 and x_2 are positive numbers, and we want to determine the values of x_1 and x_2 that maximize y subject to the limitations of the constraint equations. Our brief example obscures the fact that a realistic complex problem requires simultaneous consideration of literally hundreds of factors and constraints.

The strategy for the search by the linear programming for the optimum values of x_1 and x_2 has three parts.

1. Locate a set of values at the extreme point of their range. In the example, when x_1 is as small as it can be, that is, 1, x_2 cannot be larger than 3. When x_2 is 1, x_1 cannot be larger than 2.
2. Compute the rate of change of y between these extreme solutions, and choose the largest rate as the indicator of the direction in which the solution should be changed. Go to the associated extreme point.
3. Repeat the previous step until a point is found from which the rate of change of y to all other points is equal to or less than zero. This

point specifies optimum values for x_1 and x_2.

The common-sensical approach of linear programming which is implemented in these three steps is to locate extreme values of feasible solutions and let these extreme values define an area within which the optimum solution lies. The area is searched for the optimum values by monitoring the rate of change of the magnitude we want to maximize. When we find the point where the slope is zero and no further increase is possible, we have our optimum values.

The process we have just described would be the method most people would use intuitively to solve a less complex problem by hand. For a problem with many variables the high speed of the digital computer allows us to use the same approach instead of resorting to more sophisticated mathematics. The many small and repetitive calculations necessary would not be feasible to calculate without the computer. In addition, the large memory of the machine permits the storage of the details of a large problem with hundreds of variables and constraints. In linear programming the computer has replaced the need for a more powerful methodology for dealing with the problem.

The reader who would like to know more about mathematical programming techniques is directed to references (5–8) at the end of the chapter. Another use of linear programming is in the management of large-scale projects. The issue to be determined is what sequence of operations will enhance the completion of the project. Means of minimizing delays and optimally reaching the project target are given by two methods: CPM, critical-path methods, and PERT, program evaluation and review technique. An extensive literature exists on these topics.

Sensitivity Theory

From the cateogry named Theory of System Structure, we want to review the basic concepts of *sensitivity analysis*. This topic is a natural complement to the linear programming just discussed. When we wrote the equation for our truck-fleet example,

$$y = 5x_1 + 2x_2,$$

we assumed that we knew the integers 5 and 2 as the parameters identifying the capacities of the two types of trucks. Parameters of a complex system are seldom known with precision. Therefore, in accomplishing the optimization of the linear programming we must be aware constantly of how sensitive the final solution is to small changes in the parameters of the system. A "sensitive" process is one in which very small changes in one or more parameters markedly affect the outcome of the process.

In the design case of the truck fleet, if a sensitivity analysis revealed that a change in the capacity of the small truck from 2 to 3 tons increased the value of y in a significant way, this information would influence our design. In the countercase of analysis of a complex phenomenon, the parameters identified by sensitivity analysis as extremely influential must be carefully measured and any uncertainty in their values taken into account.

This description of sensitivity analysis shows its most basic feature to be caluclation of the rate of change of the magnitude of the value to be maximized (or minimized) when one parameter is changed. Mathematically, partial differentiation is immediately implied as a key tool.

Before being more specific about implementing sensitivity methodology, we will list a representative set of parameters that should be considered in our truck-fleet example:

truck capacity	number of loading crews
type of truck	crew time available
number of loading docks	cost of trucks
cost of crews	budget limitations

These parameters are dimensions of the structure of our resource-allocation problem. Three types of dimensions appear in the list. The first type influences the contribution of the x factors to y, and truck capacity is an example of a "productivity factor." The second type of parameter, designated a "specification factor," is exemplified by the number of loading crews available. This parameter constrains the x factors. The last set of dimensions are those affected by operating within a fixed budget. The cost of each truck is an example. Each of these types of constraints requires a somewhat different approach to analyzing and minimizing sensitivity. In general, the approach is to superimpose changes in y due to changes in individual parameters. This evaluation of a linear-sensitivity variation is discussed in detail in references (*9*) and (*10*).

A fourth type of sensitivity computation, called *breakeven analysis,* is nonlinear and will be reviewed to conclude this section. Implicit in this and other expositions of complex computations is the fact that we are forced to write *about* them, since the mathematical level required to *do* the computations cannot be expected of all or even many readers.

Breakeven sensitivity analysis is aimed at comparing competing systems with regard to the point at which benefits exceed expenses. Both the expenses and benefits will have fixed and variable components. The variable parameters will be tested under nonlinear conditions to determine their effect on the breakeven point. This method is especially pertinent to large-scale urban systems which have substantial fixed investments and costs, and the issue to be determined is the level of system activity, a measure of benefits, that is necessary to determine desirable system performance (2).

If the issue is the design of a large urban rapid-transit system, it is extremely difficult to predict what future passenger traffic will be. Further, the benefits that the associated urban area will accrue are almost impossible to estimate except as reflected in the system passenger load. What we can do is evaluate competing systems designs on the basis of a predicted volume of service by computing fixed and variable operating costs to determine which system has the best breakeven point.

Consider the case of two competing urban transit designs. The first has the smaller fixed cost to establish the system, and the second has a lower operating cost. Designating the systems S_1 and S_2, we can represent their costs as being composed of fixed and variable factors

$$S_1 = C_1 + O_1,$$

$$S_2 = C_2 + O_2,$$

where C and O represent the dollar volume of fixed costs and operating costs, respectively. Simple representative plots of these data are shown in Figure 2-3. Each curve starts at the fixed cost and increases according to the operating expenses. Based on the assumptions, S_2 is the best system prior to the curve crossover point; S_1, after the breakeven point. In this extremely simplified example we can vary now some of the transit system design parameters and observe the shift of the breakeven point. Typical tradeoffs that can be evaluated in this manner would be the choice between smaller, less expensive vehicles operating less frequently. Clearly, volume of service is only one aspect of benefit or the lack of it, and in a realistic problem other dimensions which can be identified and quantified should be included in the sensitivity analysis.

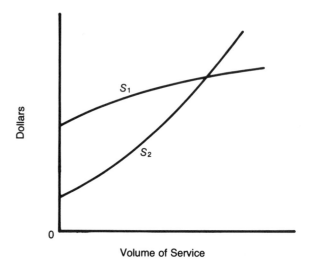

Figure 2-3

In a nutshell, then, sensitivity analysis allows us to deal with uncertainty in the values of the parameters while optimizing the solution to the resource-allocation problem.

Economic Systems

The use of modeling techniques in economics is generally referred to as econometrics. Regardless of the nature of the problem, or the discipline with which it may be associated, economic factors and the estimation of costs or values are implicit in every social system investigation. Issue may be taken with respect to the weighting of the economic factors, but little argument will be forthcoming regarding their inclusion. The previous sections on linear programming and sensitivity have concentrated on analytical methods·which are useful only if necessary factors, in this case economic, are available. In this section the difficulties of estimating and using these factors is examined.

In the estimation problem the researchers must interpret bookkeeping and accounting procedures in such a way that accurate data for systems analysis are available. These accounts are very carefully kept for tax and corporate reasons and possess many deficiencies from the systems viewpoint. The first difficulty is the effect that traditional accounting techniques have on determining the price paid for materials coming into the system. Nominal values are entered, but hidden in the accounting are surcharges, discounts, delivery and storage costs, insurance, special services to the provider, and possibly a schedule of payments. Level of quality of the material and hidden benefits should also be investigated.

Assuming that with great effort these factors can be extracted from the accounts, many of the values of the factors will be in error because of a traditional conservative position of understating assets. Inflation effects, depreciation procedures, and market value all influence the worth of institutional resources, and these effects may be very difficult to interpret becuase of accounting customs. In summary, then, on the estimation problem, if we know what cost we want to measure, the achievement of precision in that measurement is very difficult.

The other aspect of determining costs is knowing what is the appropriate cost to measure for your analytical needs. For systems analysis we want to know the marginal costs. *Marginal costs* are based on an optimum tradeoff between resource value and product value. Therefore, marginal costs are a description of the best way to combine resources to produce goods or services. The fundamental cost of a resource will be taken as the decrease in wealth which results from committing this resource to a particular use before any benefits that might accrue from its use are taken into account (4). This definition permits a classification of various costs so that the proper choice for analytical needs can be made. In the next six paragraphs the definitions of the various costs will be reviewed.

1. Opportunity costs: The *opportunity cost* is the value of the resource if it were not committed to the process being investigated. This cost can be very different from the *outlay cost* of the resource. If a city has or acquires land for a housing project at an arbitrary price (high or low), the opportunity cost is the market value of the land. Indirect opportunity costs include the sacrifice of future earnings of the land, or the value of the

land for a use not reflected in the market value. Obviously opportunity costs are difficult to measure, but the attempt must be made to obtain this cost of resources.

2. Fixed and variable costs: The marginal-cost concept is based on evaluating the change in the total cost due to a change in one of the resources. It is mathematically similar to evaluating the sensitivity of a system to the change in a parameter. To evaluate the change in the total cost requires knowledge of both the *fixed* and *variable costs*. These costs have clear intuitive meanings; however, the time frame is critical, because no cost is fixed for all time. The separation of fixed and variable costs must be done within a specific time span and with respect to a particular issue.

3. Long-term costs: To evaluate the change in marginal costs over variable time periods requires consideration of a number of factors. Sample pertinent factors included the level of system operation, changes in technology, economics of scale, and future opportunities. Both long-term and short-time costs require careful specification of time scale and detailed investigation into the nature of the process.

4. Average costs: A common error is to fail to differentiate between *average cost* and *marginal cost*. They can be equal only if there are no fixed costs and the supply of the resource is not markedly affected by the amount being procured. If a city is building a series of housing developments, and land is scarce, the price will continue to escalate and the average and marginal costs will differ. Recall that our goal for marginal costs is that the marginal product per unit cost of each resource must be equal for all resources. Because we are interested in large-scale urban phenomena, we do not expect that marginal and average costs will be the same.

5. Joint and traceable costs: The difficulty of identifying costs, as discussed in previous paragraphs, is clearly complex, but the difficulty is compounded by the fact that many expenses are joint costs for several different activities. Here, again, accounting practice and tradition must be understood in order to determine true marginal costs. Generally, material resources are more easily traceable than service resources, such as secretarial, managerial, and maintenance services. For years, federal monies allocated for research projects have helped support students, construction, libraries, and many other services of a university. Data provided by accountants would have to be dissected to reveal this.

6. Past and future costs: This last category of costs attempts to emphasize the fact that no costs are timeless. If our model is being used for prediction, estimates of future costs must be obtained as accurately as possible. Sources of information are available which maintain data allowing the reasonable estimate of the trends of changes in costs. If a new system is being designed, alternatives need to be analyzed in terms of how they will affect future wealth. Only a precise estimate of future costs will permit such a choice between alternatives.

The variety of cost considerations seems endless. One last point is necessary to complement our six categories of cost. The cost of money itself changes over time. The opportunity cost of money is the correct value of money to use in systems analysis. References (*11*), (*12*), and (*13*) give more detail on the economic topics of this section.

Simulation

In this section, and the four that follow, five major methodologies of systems analysis will be examined. In each section the intent will be to determine what the methodology is, what it can and cannot do for us, and we want to look at some examples of its usage. This second half of Chapter 2 completes the evolution of systems methodology as we perceive it. Early in the chapter basic vocabulary and system disciplines were reviewed so that problem classification could be done. The second level of methodology, such as sensitivity analysis, utilized combinations of basic disciplines to deal with systems features. Now even more complex combinations of basic disciplines and level two techniques form more powerful and general methodologies for systems analysis. All three levels of methodology are embedded in the long list given at the beginning of the chapter. And the candidates from each level represent a reasonable and manageable sample of every level of methodology.

But now, after that preamble, let's get on with an exploration of simulation. A rather general and practical definition is as follows. Simulation is the systematic abstraction and partial duplication of a phenomenon to effect (1) the design of a system under particular conditions for a specific purpose, (2) the analysis of the phenomenon, or (3) the transfer of training from a synthetic environment to a real environment. Our requirement is for item 2. There are a number of advantages and disadvantages to a simulation study as an analytical tool, and it is appropriate to review them at this stage (*14*).

Perhaps the most desirable features of simulation include intelligible results and a freedom from the requirement of mathematical sophistication. Complex mathematical operations can be performed via simulation computer programming by a user with only modest skills. And the mathematician himself can escape from some of the rigors of his analysis techniques. The output of a computer simulation can take one or more of several useful forms: (1) a yes or no, (2) a set of data ready for statistical analysis, (3) a graph, (4) a table, or (5) a sentence. You may have noticed how the words "programming" and "computer" started appearing soon after the topic of simulation was begun. The simulation of a phenomenon may take many forms. In years past, pilots experienced simulated flight in the well-known Link trainer. This and many others were physical representations of complex phenomena. In recent years, simulation has become synonymous with computer simulation and the computer most used has been the digital computer.

The very process of developing a program for a digital-computer simulation has inherent advantages. Assuming that any machine errors are not significant, the program requires a logical, properly timed sequence of modeled operations in order to produce a simulation output. Although there may be gaps in mathematical or verbal theories, these gaps must be filled for the program. In this way, the modeler is moved toward identifying significant features of the phenomenon.

For the advantages just reviewed, there are disadvantages. With respect to this last point on filling gaps in theories, if little information is available, the gap must be filled, if only by speculation. Therefore, in order to have an operable program, the investigator is forced to conjecture, and the representation of reality may be a poor one. The availability of languages for programming, and the organization of the computer to be used for the simulation, have substantial influence on the ease or difficulty of accomplishing the modeling process. Computer-simulation outputs can be very nearly independent of the conceptual models on which the program is based, and very dependent on the programming techniques used. For this reason, computer-simulation outputs must be carefully validated to ensure that they are a consequence of the conceptual base.

A problem solved by simulation often leaves the investigator with no clear understanding of general relationships. The simulation output results from use of particular parameters and variables, and these values may obscure the total range of output behaviors of the model.

If a simulation model becomes extremely complex, it may be as difficult to understand as the phenomenon it portrays. In any case, a credible simulation offers a rare opportunity for experimentation which could never be accomplished with the real-world process. The computer simulation will attempt to operate on any data subject to specified constraints, will forget on command its previous experimental experiences, and will run day and night if the experimenter desires and can afford it.

The issue of cost is our last disadvantage. A review of the history of large-scale simulation reveals that each has cost much more than expected, and the time of completion has been excessive.

To go to more specific matters now, these are the typical steps necessary to accomplish a computer simulation (*15*).

1. Definition of the problem.
2. Planning the study.
3. Formulation of a mathematical model.
4. Construction of a computer program for the model.
5. Validation of the model.
6. Design of experiments.
7. Execution of simulation runs and analysis of results.

These steps are an expanded and modified version of the basic steps of the systems approach given in Chapter 1. Steps 1 and 2 should produce a clear concise verbal description of the phenomenon and planning sufficient to allow accomplishment of the remaining steps. The construction of the model requires two parts: (1) establishing the structure of the model based on required performance measures, and (2) collection of data to provide model parameters. Construction of the computer program must be considered while forming the model, since the program is severely influenced by the nature of the model. If the two are compatible, the actual programming is a well-defined task. The programming may be difficult and take a long time, but the means and skills are normally available.

Validation of the model is important, requires a good deal of judgment, and may be difficult to do. The simulation can be run on the computer to see if the model behaves as planned. Conceptual errors in the model must be separated from programming errors and each corrected. Historically verifiable data must be input and output of the model as a means of verifying the validity of the simulation.

The experiments of step 6 are designed to meet the objectives of the simulation study. Specific questions have been asked, particular data are needed, and the experimental computer runs must be made within the budget and time constraints. In the last step when the simulation runs have been accomplished, the answers to the study objectives can be formulated. Two measures of success of a simulation will be how directly the results bear on the study issues, and how well extraneous data are minimized.

The simulation example we shall present in this section will be an "industrial dynamics study" of the type credited to J. W. Forrester and his colleagues at the Massachusetts Institute of Technology (*16*). The general objective of this type of study is to determine how the organization of an enterprise will affect its performance. The simulation model treats the system as a continuous process demonstrating its characteristics, and then alternative policies are tested on the model.

First, a concise statement of the phenomenon: We want to investigate the rate of sales for the manufacturer of air conditioners to the new-house market. Housing contractors have observed that the rate of selling houses depends on the number of families who do not yet have a house. That is, when the market place has few potential homeowners, rates of sales are low; and the counter is also true. In the same way, rates of sales of air conditioners decline when new house sales decline and vice versa. This is a

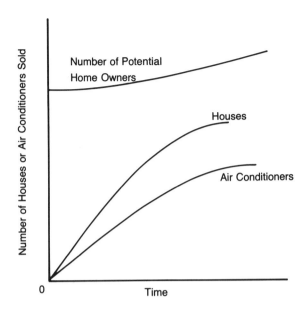

Figure 2-4

simplistic description of a portion of a complex phenomenon, but it is manageable for our purposes. The description indicates the data we should take, and the data are plotted in Figure 2-4. The upper curve plots the data for the time-varying number of potential homeowners and the curve is designated $z(t)$. The time sales of houses is denoted $y(t)$, and $x(t)$ is the time sales of air conditioners. With the description of the interaction, and these data available, the modeler decides that a mathematical model is feasible and will be useful. The trend of the slowdown of sales as the housing market becomes saturated can be represented by the equation

$$\frac{dy}{dt} = C_1(z-y).$$

The term dy/dt represents the rate of home sales, and the term $(z-y)$ is the number of families without homes. The equation relates these two variables through a proportional constant, C_1.

In the same way the equation

$$\frac{dx}{dt} = C_2(y-x)$$

relates the rate of sales of air conditioners, dx/dt, to the number of homes without air conditioners, $(y-x)$. Substituting one equation into the other so that the combined effect of the two trends can be seen gives

$$\frac{d^2x}{dt^2} + (C_1 + C_2)\frac{dx}{dt} + C_1 C_2 x(t) = C_1 C_2 z(t).$$

The only new-looking term in the combined equation is the first one, which depicts the rate of change of the sales rate of air conditioners. This last equation can be identified as a second-order linear differential equation, and its solution is well known. The equation describes the functioning of many phenomena, including a diving board, a bassoon, and the ecology of the lemming.

The computer model of the second-order equation can now be programmed so that for values of C_1, C_2, and $z(t)$, the sales of air conditioners as a function of time, $x(t)$, can be determined. Figure 2-5 gives three possible plots of $x(t)$ that we would receive at the output of the computer simulation. The horizontal line represents the housing demand. Curve A depicts the case where sales of air conditioners start at a rate which slowly decreases to zero when all houses

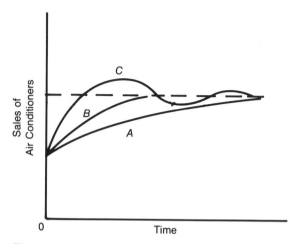

Figure 2-5

are being supplied just as needed by a constant number of sales. Curve B is similar to A, except the initial sales rate is higher, and the time required for a steady number of sales is shorter. Curve C shows the highest sales rate, and the curve rises so quickly that it "overshoots" the demand. This could be interpreted as the stockpiling of air conditioners by the contractor until the houses are finished, and then oscillating between being oversupplied and undersupplied until steady sales are reached. The actual curve in a real situation will depend on the values of C_1 and C_2. The computer simulation will allow the testing of a variety of values of C_1 and C_2. Another interesting experiment that could be done is to investigate the effect of the sudden lowering of the cost of a home mortgage. There would be an immediate jump in the curve $z(t)$, potential homeowners, and the two sales response curves would be affected. The variation of C_1, C_2, and

$z(t)$ would allow the simulation of a variety of conditions, and could lead to useful recommendations to the contractor and the air-conditioning manufacturer.

The accomplishment of the seven steps of simulation included validation of the mathematical model by observing historically verifiable inputs and outputs, and then trying new inputs so that predictions could be made regarding the consequences of the new actions.

Planning–Programming–Budgeting System

The planning–programming–budgeting system (PPBS) is the second complex systems methodology we want to examine. PPBS is a systems approach to administrative planning, and in the mid-1960s a Presidential pronouncement incorporated program budgeting into the entire federal structure (*17*). Conceptually, however, the nature of PPBS is such that, with care, it can be applied to nearly any type of human organization. The intent of PPBS is to provide a more rational basis for the allocation of scarce resources among competing programs. The primary departure from traditional planning procedures is the shift of the focus to the outputs of an organization, rather than the inputs. In terms of budgeting, the shift is from justification to analysis. That is, budget decisions are influenced by explicit statements of objectives and by a formal weighing of the costs and benefits of alternatives. In this way performance is measured against objectives. This is a move away from intuitive processes based on experience and hunch as a basis of decision making.

To set the scene for the details of PPBS, it is useful to briefly review some of the points related to a more traditional budget approach. In the past, budget has been organized according to the subdivisions of the organization. The particular subdivision, say a department, would project a year ahead its needs for resources to cover expenditure activity for personnel, construction, supplies and equipment, or whatever was appropriate to the departmental task. The last sentence describes an input-oriented system which emphasizes bringing together resources to carry out the organization's programs. For a large organization, such as the federal government, the name and mission of a subdivision such as an agency does not really say what the agency does. The personnel content may not be specifically informative, and relationships with other agencies are often obscure. A yearly budget, with difficult to assess long-term effects, is not informative with respect to financial decisions. The annual budget does not tell us whether or not program objectives are being met or whether spending levels are appropriate. The PPBS approach is: "A budget should be a financial expression of a program plan. Setting goals, defining objectives, and developing planned programs for achieving those objectives are important integral parts of preparing and justifying a budget submission" (*17*).

From the general description of PPBS just given, it may be realized that PPBS is a detailed expression of systems analysis. With this relationship come many of the difficulties of the systems approach. It is not likely that the entire output of an organization can be quantified and measured. The methodology cannot substitute for good management or permit organizations with insufficient resources to achieve their objectives. Nor is this approach to budgeting a systematic economic measure which sacrifices goal accomplishment to minimize costs.

That is what PPBS is not. PPBS is a methodology for the total budget-making process so that resources are allocated efficiently to achieve specified objectives. The six major characteristics of PPBS are analytic modes, planning, programming, budgeting, structural cohesion, and administrative lines of action (*18*).

1. Analytic features: Just as in the basic systems approach, analysis is used to determine and evaluate alternative courses of action based on their cost and benefit to the organization. As much as possible, benefits are quantified. Analysis provides organizational policy appraisal rather than budget justification.
2. Planning: Planning is the process of making multiyear projections of budgets, objectives, and future conditions. In this way the outcomes for various courses of action can be systematically evaluated.
3. Programming: Utilizing the information from analysis and planning, goals and alternative programs can be related through specific lines of action. Procedures and data are prepared, manpower is assigned, and the material and facilities necessary for the program are allocated.
4. Budgeting: This key step of PPBS relates programs and resources which are then transformed into budget dollars for future years. The relationship between dollars and output or programs is critical. Each subdivi-

sion of the organization is identified as providing a part of the output program.

5. Structural features: Programs must be analyzed in terms of their output characteristics. The composite of the output activities defines the total work of the organization. Some classification scheme is used to categorize the several possible levels of program outputs.
6. Administrative features: From the administrative viewpoint, the nature and form of PPBS has the potential for providing decision makers with a complete basis for rational choice. Discrepancies in concerns for human values are more easily detected.

Many other PPBS characteristics exist. Objectives must be explicit in operational terms. Evaluation of objectives is done using performance indicators, if possible. System science methods can be included in PPBS to make it sensitive to human needs. The approach is basically an economic concept intended to function in the political arena. When the best features of budgeting procedures of the past are preserved and combined with the concepts of PPBS, there is a modest record of success in federal agencies, municipal and state government, and industry.

The complexity of PPBS procedures does not permit a simple expository example; however, for the interested reader there is a large literature on PPBS. Chapter 5 of reference (*18*) is an interesting survey of a number of recent projects in PPBS development as applied to educational organizations.

At several points in this section some of the difficulties associated with PPBS have been given. To conclude the section (unfortunately on a cautious note) the following concise list includes many limitations, cautions, and problems that may be experienced in applying PPBS.

1. PPBS is influenced by political considerations.
2. Societal and organizational goals are often unagreed upon and hard to define.
3. Relating means to ends is very difficult.
4. The separation of factual and value elements and short-run and long-run effects is impossible (*19*).
5. Relating specific programs to resource requirements and resource inputs to budget dollars is very complex.
6. In all but the simplest cases, benefit evaluation is difficult.
7. Evaluative criteria are not available, not agreed upon, or need refinement.
8. Use of PPBS requires skilled staff and additional resources.
9. There is no certain method of making predictions.
10. PPBS conceptually is the antithesis of traditional budgetary processes.
11. A side effect of PPBS is likely to be a centralization of authority.
12. Goal-oriented expenditures are likely to be more conservative, with a bias against high-risk programs.

Gaming

Gaming is a form of simulation that has a long history of usage as a tool for policy-making professionals. The extensive history of gaming is concentrated in *operational gaming,* particularly in the form of war games. There are indications that war gaming dates back over 3000 years, and it is likely that chess is one of the earliest versions of war games (*20*).

During the last 15 years there has been a proliferation of gaming, not only in operational devices, but for use in teaching and experimental work. The increased usage of gaming has been stimulated by the possibilities for game enrichment offered by the digital computer.

But what is the value that gaming offers us? The answer is: a systematic way of dealing with complexity—an answer that should be familiar in the light of many topics discussed thus far. For example, the behavioral scientist studying the interactions of two people through a single experimental game soon becomes overwhelmed by the thousands of interesting hypotheses or conjectures that might be tested. The game has become a vehicle for allowing a *realistic* interaction to occur in a *simulated* environment over which some *control* may be exercised.

First, then, a brief review of the uses and scope of gaming (*21*). Gaming applied in teaching has been used for teaching principles, learning skills, and portraying case histories. One of the largest contemporary uses of teaching has been in business games. In a business game relevant features of the organization and its environment are simulated, and the business man is required to experience activities in a few hours or days that would normally be spread over several years. Clearly, this can be a significant learning experience for him.

The uses of gaming for operations have been directed to "dry runs," exploration, and planning.

The "dry run" or "dress rehearsal" is familiar in military, political, diplomatic, and economic settings. Contributions to both exploration and planning can be achieved by allowing the appropriate cast of characters to interact in a rich environment with an unstructured setting.

Experimental gaming has contributed primarily to psychology and economics. Reference (22) is an example of a book in gaming in economics.

In all categories of gaming, the use of the digital computer is more and more common. The factors that determine the validity of a computerized game are:

1. Costs and availability of computers and skilled personnel.
2. Importance of computational error.
3. Administration and paper-handling costs.
4. Importance of time delays.
5. Displays and input–output instrumentation.
6. Need to automate analysis.
7. Importance of parameter control.
8. Mobility.

Our last point prior to consideration of a gaming example is with regard to game construction. In spite of the long history of gaming, the number of skilled professionals is small, and the high level of activity is a recent phenomenon. Perhaps the best way to learn the concepts of gaming is through the construction of a game. The construction steps are closely related to those of the process of model building, and to systems analysis in general. Game construction requires four skills: (1) a deep knowledge of the phenomenon being investigated so that a precise abstraction can be formulated; (2) the ability to construct a model that can be controlled, manipulated, and analyzed; (3) digital-computer programming skills; and (4) the capability of designing suitable experiments for the game and arranging the output so that the results are appropriately displayed.

Build

When we want to use gaming in the urban setting, the focus on existing urban games has been on technical decision making about the development and expansion of a real or hypothetical city. Recently, urban gaming has moved from the managerial problems to try to consider political, social, psychological, and cultural aspects. Our urban gaming example, *Build,* was formed to satisfy these goals (23).

1. To allow maximum expression of value positions by participants through resolution of intense, task-oriented conflicts.
2. To heuristically gather information on both the technical and social functioning of the city through feedback from participants.
3. To provide community participants with access to technical expertise in urban decision making, and to expose professionals to the value positions of the community.
4. To lay the groundwork for eventual development of an actual policy-making tool.

The participant roles that play Build are the mayor, zoning and city planning, school board, health and welfare department, police, national business, local business planner, agitator, and the people (parents and labor). The role for each of these participants is specified. For example, the mayor must set tax rates and department budgets for each year, be responsible for all borrowing decisions, and is able to change the tax rate on any particular business property. Each other participant person or group has a specified role.

Each participant has an input–output computer format, which includes working data, input, and available output data. The input–output format is summarized next for the mayor.

Working Data for Mayor

Estimated tax bases for year 2002
Estimated total expected taxable income is $150 million
Estimated total sales expected is $50 million
Estimated total taxable property value is $750 million
Estimated income for next year
    $12 million from income tax
    $300 thousand from sales tax
    $23 million from real estate tax

Actual expenses for year 2001 were	40 million
Actual income was	41 million
Surplus for year is	1 million

Social Parameters for Mayor (requested for year 2001)

Crime rate is 18 crimes/1000 people
A total of 38,368 crimes were reported
School dropout rate is 19 percent
School truancy rate is 8 percent
Average education level is 9.1 years
Per capita income is $740
Unemployment rate is 20.1 percent
60 percent of people are registered to vote
30 percent actually did vote
Median age is 22 years
Average family size is 5.6 people

Input from Mayor

Input tax rates for year 2002 are ?
Sales tax rate is ?
Income tax rate is ?
School board budget is ?
Police budget is ?
Probable deficit for next year ?
Any properties whose tax rate is to change ?

Each role in Build will have an input–output format such as has been summarized for the mayor. Next, an initial scenario is given to provide a setting through which the various roles can interact. The sample scenario provided by Orlando and Pennington is:

River City is a four square mile section of one of the oldest East Coast metropolitan areas. It is bounded by a river along which are found a large number of rundown factories and warehouses. The southern portion is what is often called a slum, although the people who live there are proud of their homes, and intensely interested in improving conditions. The north west corner is a stable middle class community of long standing. River City is now deeply troubled over three major issues: defacto segregation of schools, a proposed northwest-southeast expressway, and a proposed industrial park in the southwest corner. In addition, a large block of federal money has just become available for redevelopment. As the game begins, the mayor is under pressure to develop a program which can unify the community.

Build is stored in a digital computer and can be played from any telephone using a teletype unit and an acoustic coupler.

To conclude this section on gaming it is appropriate to say that though operational games are highly effective teaching and communication devices, their utility for predication and replication tends to be low. Not until the game is linked to very detailed mathematical models is this likely to be changed (*24*). In addition, with some professional groups, gaming and simulation are thought to be modern, fashionable, and avant garde. However, unless the requisite skills and effort are available, the methodologies will not be helpful (*25*).

Cost-Effectiveness Analysis

The beginnings of the analytical concept known as *cost effectiveness* are related to the evolution of systems analysis as described in Chapter 1. Implicit in the systems approach is evaluation, and, starting at the time of World War II, methodological evaluations were emerging as a part of systems analysis. The older, more general term "cost benefit" was sometimes applied, but later, the use of the term "cost-effectiveness analysis" (CEA) became more common. The Operations Research Society of America has a Cost Effectiveness Section, and one can find entire volumes devoted to CEA (*26*).

The notion of relating costs and effectiveness is intuitively desirable and has been going on for a long time. What is different today is the availability of techniques from systems analysis to determine CE, and the acceptance and use of CE at many levels and in diverse settings, as an aid to decision making. The particular decision-making process of interest is that of making a choice among possible alternatives to achieve a specific objective. As a practical matter, then, we may be interested in minimizing cost, maximizing effectiveness (which may not be easily measured), or both. Immediately implied are the specification of objectives, identification of alternatives, acceptance of performance measures, and the understanding of the influence of nondollar costs.

If the issue that a school board must decide is the location of a new grammar school, this analysis is relatively narrowly defined. There is a specific district, available land at particular costs, population distribution, existing facilities, and so on. The alternatives are similar. However, if the school board's concern is to improve the quality of education in their district, the alternatives will be very different: facilities, finances, techniques, and so on. When the alternatives are so dissimilar, the criteria for choice are far more difficult, and it is to this second problem that CE is primarily aimed.

The general process of CE we have just described is depicted by Quade in reference (*26*). Figure 2-6 portrays the salient features of CE. The general description of CE thus far and Figure 2-6 identify the elements of CE as objectives, alternatives, models, effectiveness, costs, and criteria. The first three elements should be familiar from previous discussion of the systems approach, and marginal costs were reviewed earlier in this chapter. Although most costs can be measured in money, the true cost is in terms of opportunity lost, and this will have to be estimated. The new important ingredient is effectiveness, and we will have more to say about that. But first, the neatness of the structure in Figure 2-6 is misleading. Problems and difficulties abound. A complete list of alternatives is difficult to form, and often alternatives are not adequate to attain the objectives. Models are difficult to form and measures of effectiveness are only approximate. Uncertainties throughout the process may under-

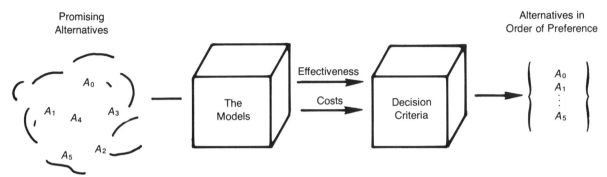

Figure 2-6

mine the confidence in the quality of predictions. The analysis, as always, must be of high quality and must be cyclically repeated to sharpen objectives, identify other alternatives, collect more data, form new models, and improve cost and effectiveness evaluations until the promise of CE is realized.

The new ingredient of effectiveness will receive our attention for the remainder of this section, after a brief look at cost. The output-oriented costing mechanism of PPBS is what we would like to use here. Establishment of such costs in the past has resulted from a mixture of political, administrative, and professional judgment. Prior to PPBS a *cost-benefit analysis* was used which attempted to quantify both costs and benefits in dollars. Reference (*27*) presents a good survey of cost-benefit analysis.

It is not easy to tie costs explicitly to the outputs for which they are incurred. Municipal budgets usually consist of expenditures and appropriations by agencies subdivided into the objects of expenditure. Since the objects are categories of purchases, the focus is on the input. In this case, the way that costs contribute to objectives is clouded. Since the municipality provides primarily services, the labor cost will be dominant and a review of other purchases uninformative. The question then becomes how the activity of those services, including managerial, contributes to objectives.

Another interesting costing quandary is that the capital costs of components of urban service systems such as branch libraries or fire substations often amount to no more than two or three years' operating costs (*28*). Capital savings strategies are likely to adversely affect operations.

The idea of measuring effectiveness, although imperfectly defined, is useful. Its intention is to tie performance characteristics of alternative systems to alternative ways of describing what they are supposed to do. Effectiveness measures are rare and performance measures or standards are used instead. The relationship between the case load of a welfare worker, or the citations issued by a policeman, and the objective to which the work is directed is dubious. Here, again, the line between an organization's activities and objectives becomes hazy.

Teitz in reference (*28*) gives an interesting example of measures of effectiveness applied to a Job Corps manpower-retraining program. If we use the dropout rate as a measure of program effectiveness, a low rate is likely to restrict program-admission requirements, blocking very worthy candidates. If entering and leaving test scores were used to determine effectiveness, long-term benefits of the program would be omitted. To use the number of participants as a factor will only minimize the program value per individual. What we are after is a measure of the effectiveness of the program for reintegrating people into the mainstream of economic and social life. The other suggested measures are only a part of this basic issue. If all effectiveness criteria cannot be combined, skill and judgment must be used to determine the relationship of each to cost.

The idea of cost effectiveness is critical to the application of the systems approach in urban settings.

Decision Making

In this final section of our review of a sampling of systems methodologies, we will, for the last time, give a short exposition of a topic more appropriately presented between hard covers, as in reference (29). Indications have been substantial that decision making at many levels is an essential ingredient of complex social problem solving. The issue of providing better information to decision makers has occurred again and again.

We will draw on a body of knowledge, decision theory, which provides a rational framework for choosing between alternative courses of action when the consequences resulting from this choice are imperfectly known (30). We want to use all available information to give us the best possible logical results. Our attitude is that we will use this methodology, determine the decision, and make the results available to the decision maker as yet another special form of information to be helpful to him and contribute to the quality of his decision making.

Decision theory is a formalization of common sense and uses two dimensions to model decision making. *Utility theory* will be used to describe *value* and *probability theory* to describe *information*. Conceptually all decisions are similar, and we will use a simple example to show the evaluation of the dimensions of decision making.

Consider the situation of the chairman of a school board who has a board split evenly on the issue of whether or not to grant pay raises to the teachers. He knows his vote will be the deciding factor, and simultaneously the teachers are meeting to decide whether or not to strike over the pay issue. The chairman must decide about the pay raise while he is uncertain about the outcome of the strike vote. His decision can be depicted in the *payoff matrix* of Figure 2-7. An alternative way of portraying the chairman's decision is with a *decision tree,* as in Figure 2-8.

As a first step, utility theory will be applied to assign numerical values to the possible outcomes. The basic principle is that if a decision maker is indifferent between two alternatives, the expected utility (value) of the alternatives is the same. Von Neumann and Morgenstern proposed that equally attractive alternatives could be detected by careful interrogation (31). For example, suppose that you had the choice of these two alternatives: (1) a tax-free prize of $1 million,

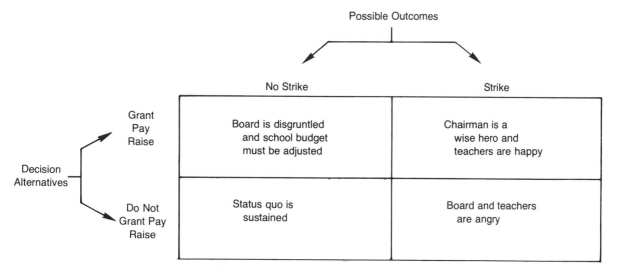

Figure 2-7

or (2) tossing a fair coin and if heads came up you get nothing, if tails appears you receive a prize of $2 million. Most people would choose the first alternative; however, it is possible to increase the magnitude of the prize of the second alternative until the choice would be very difficult to make. At this point of indifference, the value of the alternatives is equal.

In our simple school board example there are four possible outcomes. Arbitrarily assigning a value of 1 to what is clearly the most preferred outcome, that the chairman is a hero, and a value of 0 to the worst outcome, with everyone angry, we have established *our reference* outcomes. Then using interrogation we find the value of status quo as 0.91 when the chairman is indifferent to the outcome of status quo and a 10:1 chance of being a hero as opposed to everyone

angry. In the same way, the value of the disgruntled board is measured at 0.667. For the reader not familiar with the quantification of preferences, these numbers may appear very mysterious. Many assumptions of utility theory have not been discussed; however, the method has proved to be of practical value, and further reading is recommended.

Next we turn our attention to determining a way to evaluate which outcome is likely to occur, that is, strike or no strike. We are certain that there will be a strike vote, but not certain which way the vote will go. Common practice is to assign 1 to events that are certain to occur, and numbers less than 1 when the probability of occurrence of the event is uncertain. If the event cannot occur, 0 is assigned. The use of these number assignments called probabilities is familiar to most peo-

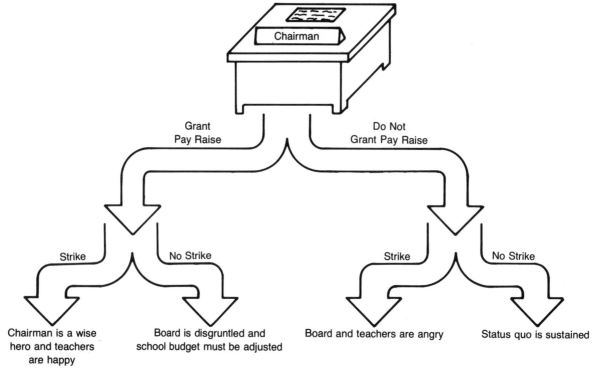

Grant
Pay Raise

Do Not
Grant Pay Raise

Strike   No Strike   Strike   No Strike

Chairman is a wise   Board is disgruntled and   Board and teachers are angry   Status quo is sustained
hero and teachers   school budget must be adjusted
are happy

Figure 2-8

ple. If the strike vote is equally likely to go either way, someone might express this as a "50:50 chance." What we want here is *subjective probability,* that is, in the judgment of the chairman of the school board, what the likelihood of a strike is. When we interrogate him he responds "2:1, there will not be a strike." Interpreting this answer in the 0 to 1 range, the probability of a strike is 0.333, and the probability of no strike is 0.667. Figure 2-9 gives the payoff matrix with all quantified entries present.

Finally, we want to use the quantified utilities and judgments to indicate the preferred decision.

A decision rule is needed, and a common-sense selection of a rule would be to make the decision which maximizes the utility. Summing the probability and value of the outcomes along the rows gives

grant pay raise $= (\frac{2}{3})(\frac{2}{3}) + (\frac{1}{3})(1) = 0.777,$

do not grant
pay raise $= (0.667)(0.91) + (0.333)(0)$
$= 0.607.$

The payoff matrix, subject to the assumptions, suggests that the school board chairman would

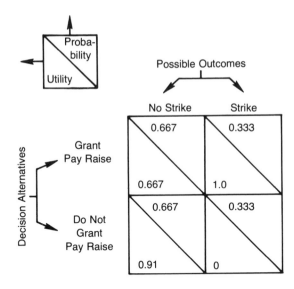

Figure 2-9

be consistent with the data if he voted for a pay raise in our very simplified example.

The running example of this section has displayed the conceptual thread common to many decisions, however complex. Note that the ingredients of the information supplied to the decision maker are very visible, as expected in the systems approach.

References

(1) R. F. Vidale, University Programs in Systems Engineering, *IEEE Trans. Systems Sci. Cybernetics,* **SSC-6**(3), 217–228 (1970).

(2) G. R. Cooper and C. D. McGillem, *Methods of Signal and System Analysis,* Holt, Rinehart and Winston, Inc., New York, 1967.

(3) V. C. Hare, Jr., *Systems Analysis: A Diagnostic Approach,* Harcourt Brace Jovanovich, Inc., New York, 1967.

(4) R. de Neufville and J. Stafford, *Systems Analysis for Engineers and Managers,* McGraw-Hill Book Company, New York, 1971.

(5) G. B. Dantzig, *Linear Programming and Extensions,* Princeton University Press, Princeton, N.J., 1963.

(6) S. I. Gass, *Linear Programming Methods and Applications,* McGraw-Hill Book Company, New York, 1958.

(7) G. Hadley, *Linear Programming,* Addison-Wesley Publishing Company, Inc., Reading, Mass., 1962.

(8) M. Simonnard, *Linear Programming,* Prentice-Hall, Inc., Englewood Cliffs, N.J., 1966.

(9) F. S. Hillier and G. J. Kiekerman, *Introduction to Operations Research,* Holden-Day, Inc., San Francisco, 1967.

(10) D. J. Wilde and C. S. Beightler, *Foundations of Optimization,* Prentice-Hall, Inc., Englewood Cliffs, N.J., 1967.

(11) J. Johnston, *Statistical Cost Analysis,* McGraw-Hill Book Company, New York, 1960.

(12) S. L. Optner, *Systems Analysis for Business Management,* 2nd. ed., Prentice-Hall, Inc., Englewood Cliffs, N.J., 1968.

(13) K. Seiler III, *Introduction to Systems Cost-Effectiveness,* John Wiley & Sons, Inc., New York, 1969.

(14) J. M. Dutton and W. H. Starbuck, eds., *Computer Simulation of Human Behavior,* John Wiley & Sons, Inc., New York, 1971.

(15) G. Gordon, *System Simulation,* Prentice-Hall, Inc., Englewood Cliffs, N.J., 1969.

(16) J. W. Forrester, *Industrial Dynamics,* The MIT Press, Cambridge, Mass., 1961.

(17) D. A. Page, The Federal Planning–Programming–Budgeting System, *Am. Inst. Planners J.,* 256–259 (July 1967).

(18) H. J. Hartley, *Educational Planning–Programming–Budgeting, A Systems Approach,* Prentice-Hall, Inc., Englewood Cliffs, N.J., 1968.

(19) R. E. Millward, PPBS: Problems of Implementation, *Am. Inst. Planners J.* (Mar. 1968).

(*20*) R. D. Duke, Gaming Urban Systems, *Planning,* 293–300 (1965).

(*21*) M. Shubik, Gaming: Costs and Facilities, *Management Sci.* **14**(11), 629–660 (1968).

(*22*) S. Siegel and L. E. Fouraker, *Bargaining and Group Decision Making: Experiments in Bilateral Monopoly,* The Macmillan Company, New York, 1960.

(*23*) J. A. Orlando and A. J. Pennington, Build—A Community Development Simulation Game, taken from a manuscript prepared for the 36th National Meeting, Operations Research Society of America, Miami Beach, Fla., Nov. 10–12, 1969.

(*24*) A. G. Feldt, Operational Gaming in Planning Education, *Am. Inst. Planners J.* (Jan. 1966).

(25) R. L. Meier and R. D. Duke, Gaming Simulation for Urban Planning, *Am. Inst. Planners J.* (Jan. 1966).

(*26*) T. A. Goldman, ed., *Cost-Effectiveness Analysis,* Praeger Publishers, Inc., New York, 1967.

(*27*) A. R. Prest and R. Turvey, Cost-Benefit Analysis: A Survey, *Econ. J.,* No. 300 (Dec. 1965).

(*28*) M. B. Teitz, Cost Effectiveness: A Systems Approach to Analysis of Urban Services, *Am. Inst. Planners J.* (Sept. 1968).

(*29*) R. D. Luce and H. Raiffa, *Games and Decisions,* John Wiley & Sons, Inc., New York, 1957.

(*30*) D. W. Worth, A Tutorial Introduction to Decision Theory, *IEEE Trans. Systems Sci. Cybernetics,* **SSC-4** (3), (1968).

(*31*) J. von Neumann and O. Morgenstern, *Theory of Games and Economic Behavior,* 2nd ed., Princeton University Press, Princeton, N.J., 1947.

**Part II
Development Models**

CHAPTER 3

Planning Urban Development

Background

In recent years there have been a number of projects in which the systems approach has been used in an effort to simulate and achieve understanding of the interaction of the multitude of variables relevant to urban planning. That activity is a contemporary chapter in a long history of the efforts to develop cohesive plans for improving the housing, the industrial plant, and the aesthetic environment of cities (1). The federal government has for forty years been involved in programs designed to halt the decay of urban housing. Early versions of federal programs funded renewal of sections of the cities based on evidence of the degree of deterioration. It soon became apparent that coping with decay "hot spots" was not adequate and that general deterioration was occurring on a grand scale in our cities. Planners and others have sought to understand the urban-decay mechanisms so that a coordinated, intensive, and long-term effort could be made to restore urban virtues. Three ingredients have combined to make systems analysis feasible and useful. The data base on all aspects of the urban environment has substantially improved, the theories of urban land use are better understood, and the digital computer became capable of providing computational capacity for large-scale modeling. In other words, to a degree, the data, theory, and methodological gap common to most complex social processes has been closed.

Examples of models of urban structures will be discussed later; however, at this point it is helpful to review their general characteristics. The models have developed with very different emphasis. Some have focused on employment opportunities, others on transportation patterns to control land use, and others on housing. The urban models have not been integrated; that is, the subsystems were modeled by not one but a variety of strategies. Typically urban modeling projects cost more, took longer, and were not as useful as the participants hoped. But it still is early in the urban modeling business, and many planners recognize it as a useful tool with great promise for the future.

Systems Viewpoint of Planning

A critical ingredient in planning is the determination of objectives. Regardless of the social unit under consideration, textbooks on planning will catalog categories of objectives in lists such as short and long range, instrumental and strategic, general and specific. Frequently the classification is abstract, nonsubstantive, and not very useful. Often there is an emphasis on financial objectives which limits recognition of many realistic activities. The systems viewpoint is that a more general model of a social unit is required which is capable of dealing with the multiple dimensions of performance and structure (2). Such a model should integrate concepts from all relevant disciplines.

Gross in reference (2) proposes that it is helpful to distinguish between two kinds of performance at the output of the social unit: producing outputs of services or goods, and satisfying various interests. Breaking down these two kinds of outputs, we can list seven kinds of performance objectives as consisting of activities to

1. Satisfy the varying interests of people and groups.

2. Producing outputs of services or goods.
3. Making efficient use of inputs relative to outputs.
4. Investing in the system.
5. Acquiring resources.
6. Doing all these things in a manner that conforms with various codes of behavior.
7. Varying concepts of technical and administrative rationality.

Figure 3-1 gives a simple visualization of these factors.

The intent of the discussion thus far has been to emphasize that the first elements of both the structure and performance of the general model are people and the satisfaction of people's interest. Financial, technological, and other elements, and the decisions associated with them, are ways of thinking about people and their behavior.

Given a specific urban unit, then, the classification and language of the seven performance objectives comes alive as a language of purposefulness. Complementing the performance objectives in our general model are the structure objectives. Gross identifies these as (1) people, (2) nonhuman resources, (3) subsystems, (4) internal relations, (5) external relations, (6) values, and (7) management subsystem.

The seven structure objectives do not have the intuitive appeal of the seven performance objectives, and so a one-sentence description is listed next for each.

1. People: The structural objectives based on people relate to the type of individuals, their quality, and their quantity.
2. Nonhuman resources: Entities such as land

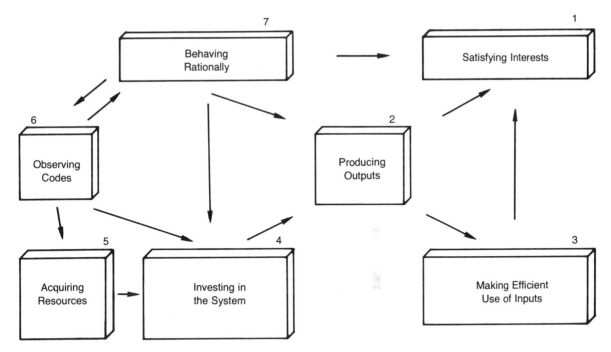

Figure 3-1

and equipment influence objectives like investments.

3. Subsystems: The subsystems that form the fine structure of the social unit are purposeful and identifiable by their role or function.
4. Internal relations: Cooperation among subsystems allows the total system to exist as intended, and cooperation is based on certain commonly accepted objectives for future performance.
5. External relations: Relationships between the urban unit and organizations in its environment may take many forms, and dealing with these external influences requires particular structural objectives.

6. Values: The pattern of values of individuals and subsystems guide aspects of the total system in the form of structural objectives.
7. Management subsystem: The nature of the management structure, for example, centralized or decentralized, must be reflected in structural objectives.

Planning may be defined as the process of developing commitments to some pattern of objectives. Now that we have a systems taxonomy of objectives, let us consider the strategies of planning to meet patterns of objectives. Again using Gross's classification, we have four strategies.

1. *The selectivity paradox:* Strategic objec-

tives can be selected rationally only if the planners are aware of the broad spectrum of possible objectives. The strategic objectives are in the form of specific sequences of actions to be taken by the social unit. These critical steps must be formed in terms of (a) requisite interest satisfactions to obtain external and internal support; (b) any possible crises; (c) impact upon preceding, coordinate, or subsequent events; and (d) long-range implications of current actions. To resolve the selectivity paradox a systems analysis should be done to provide a comprehensive background so that the range of all possible objectives is known and strategic objectives can be chosen rationally.

2. *The clarity–vagueness balance:* Precision in evolving strategic objectives is to be hoped for; however, there is a natural temporal limit associated with precise objectives. We can be clear about present and near-future goals, but there is likely to be some vagueness in long-range goals. Advantages associated with this vagueness are that future initiative will not be stifled; it is easier to adapt to changing conditions, if there is no reduction of long-term flexibility, and there is latitude for data and information to be considered and included later.

3. *Whose objectives?:* The strategic set of objectives to which we aspire belongs to whom? If we ask the constituents of an urban social unit, there will be a wide range of answers. It is likely that any single answer will be incomplete and deal only with a part of the total set of objectives. The best set of objectives will be the ones most widely accepted by all levels of the structure. This is a vaguer test of acceptability, but a more practical one.

4. *Conflict resolving and creating:* In the most realistic sense, planning is an exercise in conflict management. Gross summarized this point very well in the following statement: "Any real-life planning process may be characterized as a stream of successive compromises punctuated by frequent occasions of deadlock or avoidance and occasional victories, defeats, and integrations." Thus the selection of strategic objectives must be done with the awareness that they influence and will be influenced by a continuing environment of crisis management.

The Urban Setting

Following the general comments regarding the planning process, it is desirable now to collect some details of the urban setting in which modeling and problem solving will be done. Cities can be considered usefully to consist of systems and we need to find ways to define the subsystems. Before doing this, however, we want to summarize recent writings by an active participant in contending with big-city problems on a day-to-day basis. This particular appraisal was made by Henry Maier, mayor of Milwaukee at the time of this writing (3). These points will serve as a useful reference for a realistic look at the design of urban models. So much has been written on the troubles of the cities that the review of Maier's writings will be given as a recent representative of that literature.

1.   City problems cut across formal institutional and political boundaries and areas of responsibility.
2.   We need to adjust to the fact that urban prob-

lems are national problems because we are an urban nation.

3. City property taxes cannot take on burdens for which they were never intended.
4. A reallocation of national resources and a reform of federal, state, and local finances is needed.
5. There is a gap between political and administrative machinery for dealing with the local environment, and the social, economic, and physical forces which are constantly altering that environment.
6. In a typical city social problem, we found over 300 largely uncoordinated entities involved including city, county, state, federal, and private.
7. There are over 400 separate grant programs for the cities, and coordinating these is a serious problem in itself.
8. Since there appears to be no substantial social science knowledge for problem solving, it is imperative that limited resources be used in a precise and efficient way.
9. A national framework must be developed utilizing the systems approach to help solve the problems of the city.

These nine points represent part of the concerns of a contemporary city administrator, and they serve as an example of the complexity and interdisciplinary nature of social problems.

Urban Subsystems

Before proceeding with any modeling, it is necessary to have an adequate description of the urban systems structure in terms of subsystems.

Unfortunately there is little agreement in the literature on this point. To provide a basis for discussion we will use the urban subsystem structure of Wilson (*4*). He assumes that the subsystems can be defined from the following identified entities of the total urban structure: objects, activities, physical infrastructure, land, and policy. Wilson developed lists for each of these categories and defined the following urban subsystems:

1. Spatially aggregated population with substructure such as age or sex breakdown.
2. Spatially aggregated economic goods and services.
3. Residential locations, including the people.
4. Work places, including the people and jobs.
5. Physical infrastructure, such as buildings and transport systems.
6. Economic activity.
7. Transport, including people, goods, vehicles, and locations at each end of the trip.

These subsystems include the majority of the resources influenced by the physical planner but omit the subsystems of education, health, welfare, social activities, recreation, politics, and so on. In this chapter the emphasis is on physical subsystems; the other subsystems will be covered in later chapters. Figure 3-2 will serve to show the relationships between the urban subsystems. The figure shows in a simplistic way the relationship pathways among people, places, and processes. It is over these pathways that balances will be established or corrected in the urban structure by various migration mechanisms of population or economic activity.

Model Design Techniques

Before examining particular urban models which are the central topic of this chapter, it will be helpful to review the techniques of model design and modeling strategies which will be used. First, there are a number of modeling ground rules or, as Wilson calls them, "a check list" (*4*). These rules will facilitate the formation of an urban model.

1. The basic concerns of modeling given in Chapter 1 as part of the systems approach must be observed. The critical point is that the model must be formed with a very clear idea about the purposes of the model. The model should be designed specifically to meet these needs.

2. The variables of the model and the causal pathways should be identified. The variables which can be controlled contribute to the potential of model for planning or forecasting.

3. Categorize the variables in a way to meet the purposes of the model. For example, if the variable is population, it might be categorized by age, sex, income, occupation, and so on.

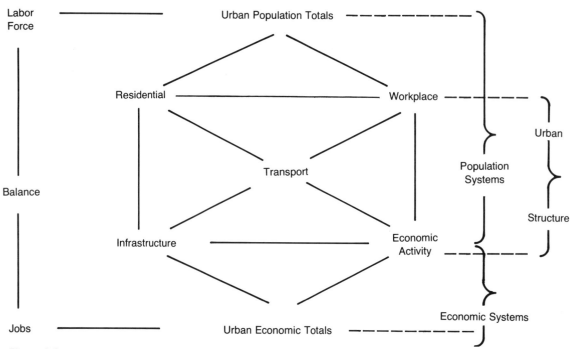

Figure 3-2

4. Determine the time frame of the model. Will the model be static or dynamic and, if so, over what time frame?
5. The theoretical base of the model must be obtained and understood. The behavioral aspects may be particularly difficult to obtain.
6. What is the available data base?
7. The model must be calibrated and tested against available data.

Using the language of systems classification, we can construct a hierarchy of models of increasing generality. This array suggests some of the logical options open to us.

1. A static model valid at one point in time with no uncertainty about the parameters (deterministic).
2. The model of option 1 but with a probabilistic (uncertain) component.
3. A dynamic deterministic model described by difference equations.
4. The model of option 3 described by a differential equation.
5. A dynamic model described by probabilistic difference equations.
6. A dynamic model described by probabilistic differential equations.

The urban modeling literature includes detailed descriptions and examples of these six models.

The Strategy of Model Design (5)

Thus far we have said something about the subsystems we want to model and the models we could consider utilizing. To be on the verge of actually producing his model, the model builder should have selected a theoretical perspective, formed a framework suitable to encompass his objectives, and postulated the existence of empirical information to permit the resolution of his problem. These three tasks are the most important ingredients of the check list of the previous section. Subtasks of these three requirements are worth discussing in more detail. The subtasks are (1) determining the level of aggregation, (2) treating time, (3) change mechanisms, and (4) solution methods.

1. The level of aggregation: When we want to model a phenomenon, there is frequently a choice of representing gross overall features, or doing a model that has fine structure and shows many details. The choices are known as *macromodeling* and *micromodeling*. The differences between the two modeling modes can be subtle in spite of the obvious differences. In urban studies macromodeling is usually associated with geography, demography, and ecology; micromodeling relates to economics and social psychology.

The macromodel depends on the statistics of large-scale behavior and collective properties. Micromodeling deals with specific resources, particular events, and individual behavior. Because of the scale of the macromodel, much is based on descriptive theoretical generalizations, and the cause–effect relations and the pathways over which they occur may be very obscure. The scope of the macromodel does not lend itself easily to financial accounting schemes. This point is a serious limitation, because it is common and useful to compare alternative programs or policies according to the single measure of dollars. Although this procedure will be repugnant to some readers, the difficulties associated with maximizing social benefits make the simplistic financial comparison necessary in many cases.

Although the fine structure of the micromodel can produce financial data very well, there are other difficulties. In economic models, for example, decisions based on rational choice hinge on the modeler's ability to characterize in great detail the preference system or relative values of the decision maker. Empirical representations of value structure are both crude in form and limited in scope.

The micromodel attempting to portray the patterns of urban development and land use must be capable of representing the total range of transactions that influence the patterns. There is such a complex mix of overlapping and interrelated markets, subtly differentiated commodities, and variations in bargaining positions that only crude micromodels of development and land-use models exist.

2. The treatment of time: Nearly all urban models describe processes whose magnitudes vary as a function of the independent variable time. Our treatment of time may enable us to determine in detail what is happening to an urban process now and then predict what will be occurring in that process at a subsequent time. There are a variety of analytical techniques available to accomplish such a temporal continuity, including comparative statics, recursive progressions, and analytical dynamics.

The method of comparative statics relies on the existence of equilibrium relationships between the exogenous and endogenous variables. These relationships ensure that internal system variables respond quickly to external changes.

The terminal condition of the system may be predicted although the exact process by which the final state is achieved is unspecified. Another use of comparative statics is to determine the state toward which the system will tend after inputting a few exogenous variables.

Analytical dynamics deals with the processes of change rather than the state of the system at a specified future date. The model takes the form of a set of differential equations and requires specification of system parameters and knowledge of the initial state of the system. All processes are endogenous except time, and the time variation of the magnitude of any variable can be plotted. If the system has no limits on the range of the magnitudes of its variables and equilibrium is not ensured, the system response may oscillate, increase substantially in magnitude, or degenerate.

The equilibrium assumptions of comparative statics and the endogenous variable requirement of analytical dynamics are frequently too demanding of the model of urban processes, and the compromise alternative utilized in the recursive progression technique allows the computation of a system's future state by using the current state and moving the changes forward in time using lagged variables. Lagged variables may be demonstrated using an interesting example of a model of the national economy (6).

Let C = consumption,
  I = investment,
  T = taxes,
  G = government expenditure,
  Y = national income.

Then the mathematical model is of the form

$$C = 20 + 0.7(Y-T),$$

$$I = 2 + 0.1Y,$$
$$T = 0.2Y,$$
$$Y = C + I + G,$$

where all quantities are expressed in billions of dollars. This is a static model and if any one quantity is given, the other four variables can be determined by solving the equations simultaneously. If the given quantity is changed, and four new variable values are calculated, our results will not show how the former values changed to the new values.

We can make this economic model dynamic by fixing a time interval, say 1 year, rearranging the equations, and identifying a lagged variable denoted by the suffix -1:

$$I = 2 + 0.1Y_{-1},$$
$$Y = 45.45 + 2.27(I + G),$$
$$T = 0.2Y,$$
$$C = 20 + 0.7(Y - T).$$

If an initial value of Y_{-1} is given, and values of G are supplied for all intervals, these four equations can be solved. Then using the computed value of Y as a new lagged variable, the computations can be repeated. Construction of a *recurrence model* is difficult, and identification of the lagged variables requires careful analysis, but the method is a useful way of dealing with time in a model.

3. Change mechanisms: The design of an urban model can be organized around *present* characteristics of the process, and this may be adequate for determining short-term changes. However, for a more serious analysis of change, a classification of variables into two categories is necessary. Lowry (5) labels these categories as *stocks* and *flows*.

Stock refers to entities sufficiently alike to be treated collectively using only the dimensions of size or number, such as acres of available land or number of residential houses. When the inventory of items in a stock changes due to additions and deletions, the time rate of the changes is the *flows*. Clearly, the time integral of a flow is the related stock. The classification of variables in this way permits the elimination of concern for exogenous variables which influence only long-range effects. In analyzing an urban retail location, if the model is organized around existing transportation facilities, merchandizing methods, and consumption patterns, the model construction will be simplified. Such a model could conceivably predict short-term changes in retail locations due to population growth but could not deal with the city's history of retail development. Many modelers use a variety of devices to try to account for changing environmental conditions with various time scales. Their explicit inclusion in the model would be very difficult.

4. Solution methods: Once the model is designed and constructed it must be operated; that is, input data must be supplied and resultant solutions received at the model output. The plan of operation of the model to achieve the processing of inputs to outputs must be a significant part of the strategy when doing the model design. Four techniques are used to provide the step-by-step operations through the model between the input and output. They are analytic methods, iterative methods, simulations, and man–machine interactions.

Analytic methods are exemplified by the recursion model of the national economy previously discussed. The equations directly relate the input and output variables, and, given the input, the model response may be computed.

If there are more unknowns than equations in the mathematical model, analytic techniques fail and iterative methods can be used. Values of some of the variables are assumed as a first approximation; then the other variables are computed analytically. Based on these results a second approximation is made and the process repeated. This continues until the values of variables converge and fail to change with further iterations.

More complex models where the scope has prevented the development of complete rigorous mathematical descriptions cannot be solved by either analytic or iterative methods. The simulation approach of Chapter 2 is more suitable in this case. The model includes an inventory of possible events, and relates the effects of the consequences of the events or the variables representing a stock or population. Exogenous events may be generated by random choice from a given frequency distribution of possibilities. The primary function of the computer is to keep an inventory of all stocks and to adjust them in response to events.

The last solution possibility involves a human participant in the simulation model of the last paragraph. In the man–machine interaction the simulation model is periodically interrupted so that the intermediate state of the system may be reviewed by a human participant. The individual may then change the results or make decisions that will affect the subsequent functioning of the model. The human operator is desirable, especially for a simulation model not capable of coping with a wide variety of unusual circumstances.

Recapitulation

A summary of the various general points of urban models will be given now before considering some of the actual models. The basic elements of any urban planning model can be stated as (7):

1. Subject: What is the model about?
2. Function: What does the model do?
3. Theory: On what theory is the model based?
4. Method: How does the model use its theory?

The most common subject matters for urban models are land use, transportation, population, and economic activity. The three basic functions of urban planning models are: *projection:* estimating the future of the subject; *allocation:* dividing the subject into subsets; *derivation:* transforming the subject by deriving another subject from it.

In projection models the future state of the subject is estimated describing the current state and then projecting the inputs through functional relationships to a state description at the end of the projection period. Reference (*8*), the Chicago Area transportation model, predicts 1980 population, manufacturing employment, and acreage by planning zone to be devoted to the major land-use classes. Such an urban model is a single interval projection, and an example of multiple interval projection is the Penn–Jersey Regional Growth Model of reference (9). This model forecasts 1980 activity but in five-year intervals and using the results of each interval to obtain the next five-year projection.

A pure allocation model is given in reference (10), the probabilistic model for residential growth. A given number of new residences are allocated to computing land zones based on a measure of attractiveness of the residences. No attention is given to projecting residential starts or deriving their sources.

Reference (11) is a combination allocation and derivation model. Examples of the transforming ability of a derivation model would be land-use classes derived from population data or traffic derived from land use. The transformation is accomplished using the model's theory of the relationship between the two subjects. Derivation occurs at a point in time and either follows projection or is associated with allocation. The model of metropolis of reference (11) has the following derivation chain. A level of industrial activity is estimated; from this the model derives the necessary residential industrial work force; after allocating the work force spatially over the plan area the required related service activities, such as food stores and repair shops, are derived. The new service functions constitute more industrial activities, requiring more work force, and so on through the cycle again. The cycle iterates until a stable spatial allocation of population is achieved which coincides with the initial level of industrial activity. Most urban models are combinations, sometimes complex, of the ideas of projection, allocation, and derivation.

One of the most difficult aspects of urban model evolution is the availability of substantive theories upon which to base the functional relationships of the model. The model may be thought of as a symbolic statement of the theory. Or if the model abstracts urban phenomena to symbolic form and relates the two structurally, theory will be developed. In general, the theories describing urban development are vague and incomplete, if they exist at all. Model builders must derive relationships as best they can in a practical fashion. Reference (7) lists an annotated bibliography of many different types of urban models.

Land-Use-Plan Design Model

The land-use-plan design model of Schlager (12) is the first urban model that we shall examine in some detail. The general intent of this model is to do comprehensive urban planning by modeling the planning process in a form amenable to computer solution. The conceptual approach to the formation of the model is based on the notion that the urban complex is a subject for design. That is, an urban plan will be evolved as "a conscious synthesis of urban form to meet human needs" (12). The conceptual alternative is to think of the urban complex as a phenomenon to be understood and explained scientifically, for example as in determining causal relationships. The design approach to planning has always been an alternative; however, the availability of high-speed computers and the work of researchers such as Alexander (13) have given a new look and at-

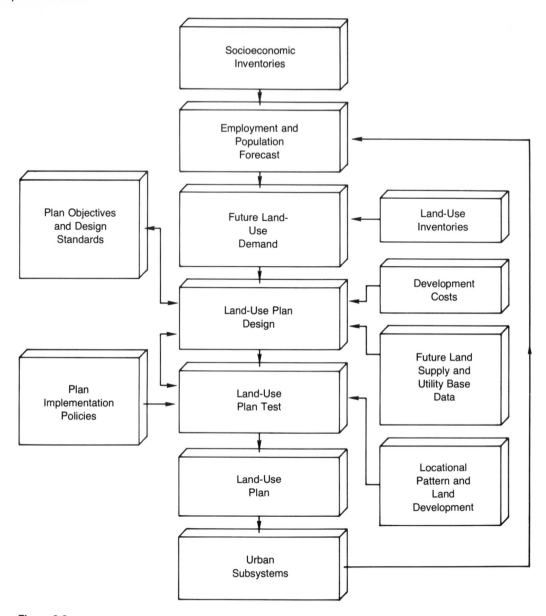

Figure 3-3

tractive potential to plan design. A block diagram is given in Figure 3-3, which shows the position of the plan design model in the planning process. The figure shows that the first process in the planning sequence is that of forecasting population and employment in order to determine future land-use requirements. Following this, the land-use requirements are specified as a detailed classified set of aggregate demands for various land uses, including residential, industrial, and commercial. These land-use demands will be one of the inputs to the plan design whose function is to provide a means of allocating scarce urban land between conflicting and competing land-use activities.

The objective of the design methodology to be used is to achieve as good as possible a match between the form and context of the land-use plan. This is difficult because the multitude of design variables interact in a complex way; however, newer forms of mathematics and the digital computer now permit effective solution of the design problem. The first requirement of the land-use plan design is to generate alternative solutions in terms of the basic measure of land use, the land itself. Three sets of variables will be used: the type of land use (quality variables), the density of land use (quantity variables), and the geographic location (location variables). These variables will be used to characterize the process of dividing the land with a grid system, and for each location the types and densities of land uses will express a measure of the activities in that area.

The second condition affecting the relationship between alternative forms and design requirements is the existence of standards that restrict the possible land-use plans. Examples of two types of land-use restrictions are exclusion of floodplain areas from development and the need to provide schools within specified distances of residential districts. The first example is a land-use standard within a grid, and the second is a restriction between grids.

In summary, then, of the design approach to a land-use plan, we want to honor any land-use restrictions and satisfy the prediction of land-use needs for urban activity by forming a land-use plan design which operates at a minimum combination of public and private costs.

The mathematical tool to be implemented is linear programming, which was discussed in Chapter 2, and the formulation of the land-use plan design model is keyed to the mathematical programming application. Recall that the task of linear programming is to maximize or minimize a function which is a combination of variables and which is subject to a set of constraints. Clearly this abstract statement relates to our interest in the land-use plan, that is, optimizing the allocation of scarce resources subject to constraints. The general form of the symbolic mathematical model will be

$$C_T = C_1X_1 + C_2X_2 + \cdots + C_nX_n,$$

where
$$C_T = \text{cost function which we want to minimize,}$$
$$X_n = \text{quality, quantity, and location variables of land use,}$$
$$C_n = \text{costs of developing the land.}$$

The constraints on the general mathematical model are of three forms:

1. The total demand requirement for each land-use category:

$$d_1x_1 + d_2x_2 + \cdots + d_nx_n = E_k,$$

where E_k = demand requirement for each land use,

             d_n = coefficients that account for land service requirements, such as streets.

There will be as many equations of this form as there are primary land-use categories, such as residential, industrial, agricultural, and recreational; and the d coefficients account for all secondary land uses, including streets and parks.

2. Limits on land uses within a grid zone:

$$X_1 + X_2 + \cdots + X_n \leq F_m,$$

where F_m is the upper limit in zone m of land use n. This constraint frequently takes the form of a density limitation.

3. Inter- and intrazonal land-use restrictions:

$$X_n \leq GX_m,$$

where G = ratio of land use n allowed relative to land use m with land uses m and n in the same or different zones. Availability of employment and shopping facilities are examples of this restriction.

Schlager reports that a land-use model for an urban area of about 30 zones requires about 60 constraint equations and 400 variables.

Four types of input data are required to operate the model:

1. Land-cost data: Land-cost and land-development-cost data must be obtained for each major land-use activity and for each type of soil by either making engineering estimates or analyzing nearby land transactions and development. Land-cost data by either means is difficult and expensive to obtain.

2. Data on the aggregate demand for all major land-use activities: This model was tested using historical aggregate land-use demand data which was available for the period 1950–1962. Future land-use needs may be forecast by applying design standards to population and employment forecasts.

3. Design standards which reflect the land-use plan objectives: Restrictions on land-use activities are available and have the effect of limiting the range of alternative land-use plans.

4. Land inventory: A land inventory that includes both land-use activities and soil characteristics is essential for the model operation.

To balance the picture it is appropriate to consider some of the disadvantages of the linear-programming land-use plan model. Perhaps the most serious limitation is the need for continuous variables for the model. Many data are collected at discrete points in time. In addition, natural discrete levels are reflected in the use of industrial parks rather than individual factory sites, and subdivision rather than individual house lots. The breach between continuous equations and discrete data can be dealt with but only at the loss of some optimality. Loss of an optimum solution occurs also when constraint relationships and design standards are nonlinear.

The design focus of the land-use model we have been reviewing allows the planner the possibility of suggesting what should be done, as opposed to explaining what is being done. The optimum design also permits the development of evaluation criteria for existing ongoing urban processes. Explaining precisely what is happening in a complex interaction is replaced by comparing the outcomes of the interaction to an optimally designed structure.

Growth-Allocation Model

The second urban model we will consider is a growth-allocation model for the Boston region (14). In this example the model will cover all aspects of urban location. A comprehensive development plan is to be evolved for the Greater Boston region, an area of approximately 2300 square miles, housing 3,400,000 people in 152 cities and towns. The model is to be operated recursively in series with a traffic model to predict development in either 5- to 10- or 15- to 20-year time intervals.

The following premises are the basis of the development of the model using theoretical reasoning and empirical testing:

1. Location of preferences of population, employment, and other activities are highly interrelated.
2. Many exogenous causal factors influence the development of land.

Using these premises and the stated approach, detailed development patterns of urban activities are predicted based on existing activities, on externally forecast regional growth, and on exogenous policy considerations concerning transportation, open space, zoning policies, public utilities, and regional growth.

The model is to be a mathematical model whose function is to evaluate the extent of interaction between activities. The model must be linear; therefore, all interrelationships must be expressed as a *summation* of the combination of the influences of the appropriate variables. The heart of the model will be the summations where any desired combination or transformation of variables will be used to describe the urban activities whose locational pattern we wish to measure and predict.

Some of the terms and definitions required to form the summation equations of the model are discussed next. The urban area of interest will be divided into small areas called *subregions,* and the model is to predict the extent of various urban activities at specified times. The urban activities are denoted *located variables.* Activities that influence the magnitude of the located variables are called *locator variables.* Using the vocabulary just presented, the following statement of theory establishes the conceptual basis of the model and supplies the information to be used in forming the mathematical equations of the model. This statement of the kernel of the theory used in the model is critical to all that follows, and it is the most impressive explicit example we have had thus far of the theoretical base of an urban model:

The change in the subregional share of a located variable in each subregion is proportional to:
1. *the change in the subregional share of all other located variables in the subregion*
2. *the change in the subregional share of a number of locator variables in the subregion*
3. *the value of the subregional shares of other locator variables (14).*

Clearly, this set of proportionality statements could be made only after the theory was hypothesized based on a careful study of existing historical data. The following equation is a symbolic form of the statement of theory:

$$\Delta R_i = \sum_{\substack{j=1 \\ j \neq i}}^{N} a_{ij} \Delta R_j + \sum_{k=1}^{M} b_{ik}(Z_k \text{ or } \Delta Z_k),$$

where Δ is a mathematical operator meaning "the linear change in"
$\sum$ is a mathematical operator meaning "the summation of"
i or j = 1, 2, . . ., i, j, . . ., N: the number of located variables,
k = 1, 2, . . ., k, . . ., M: the number of locator variables,
ΔR_i or ΔR_j = change in the level of the ith or jth located variable over the time interval,
Z_k = level of the locator variable k at the beginning of the time interval,
ΔZ_k = change in the level of the kth locator variable over the time interval,
a_{ij}, b_{ik} = coefficients expressing the interrelationships among the variables.

There are N located variables and there will be an equation of the given form for each one. We caution the reader to confirm, term by term, the relationship between the symbolic and verbal statements of the theory. The precise verbal statement, including all pertinent observations and conclusions, followed by the mathematical encoding of this theoretical information form the heart of the growth allocation model.

A computer analysis of data on urban activities for a previous period of time will give an estimation for the a and b coefficients. With knowledge of the coefficients, the equations may be solved simultaneously by substituting in each equation the pertinent values of the locator and located variables.

Now we want to be specific about the identification of the variables and the kinds of data that must be analyzed. To determine which land-use activities are related by proportional-type relationships (cause and effect), four methods of study were followed.

1. An analysis was performed to verify the relationships between subregional growth rates of population and employment and all locator variables that appeared to have any influence.
2. Changes and trends in population and employment were studied for the previous decade.
3. Variables were studied to determine their grouping effects. That is, what urban activities tended to locate close to one another or influenced the location of other activities in other settings?
4. Equations were developed which showed the relationship between population and employment growth rates and many causal urban activities which influenced population and employment locations.

Based on the studies, the located variables were identified as:

1. White collar population, including both workers and their dependents.
2. Blue collar population, including both workers and their dependents.
3. Retail and wholesale working population covered by the Massachusetts Division of Employment Security.

4. Manufacturing employment covered by the Massachusetts Divison of Employment Security.
5. All other employment.

This taxonomy of population included groups with distinct residential preferences for different subregions. This, of course, was the objective of the studies which defined the located variables.

The locator variables included many exogenous variables, and three sets of locator variables were identified.

A. Intensities of land use, zoning practices, automobile accessibilities.
B. Same as A, and transit accessibilities.
C. Same as B, quality of water service, quality of sewage-disposal service.

The model was refined, coefficients were determined, and the calibration of the model was established by using the data of 1950 to predict the subregional development of 1960. The model coefficients were adjusted so that there was minimum error between the model predictions and the actual 1960 data. Coefficients were determined for each of the three sets of locator variables based on a division of the Boston area into 29 subregions. The accuracy of predictions for the calibration run were very high. The numerical indices used to evaluate the differences between observed and predicted 1960 subregional values were the root-mean-square error, the root-mean-square error ratio, and the coefficient of determination. These indices are defined by Hill, and their evaluation indicates the high level of quality to which the growth-allocation model was developed.

Errors associated with predictions increase as the length of the period of prediction increases. With this caution in mind the model was used to predict population and employment distributions every 10 years from 1960 to 2000. In addition, 10-year recursive predictions were made where the results of each 10-year forecast were used as a base for the next forecast. Eventually the recursive predictions will be superior because they are more sensitive to changes in policy measures, land utilization, and accessibility characteristics. These effects are updated during each recursion, and the single forecast projection is not.

At the time of Hill's description of the model, it was termed "practical and operational." This evaluation indicates that the model is available to the planner, among others, and will allow the simulation of a chain of events in urban development. We could, for example, begin with variables such as the transportation system and zoning regulations and end with a desired pattern of residential and industrial development.

Future plans for this model include increasing the number of variables and subregions, and then attempting to forecast more complex urban developments such as a total multimode transportation system.

Pittsburgh Urban-Renewal Simulation Mode (15)

In this, the third urban model we are examining, the model is a digital simulation model to be used for decision-making purposes in the urban-renewal process. We are not interested in, or even able to go into the details of, the computer programming and simulation. Instead, we shall

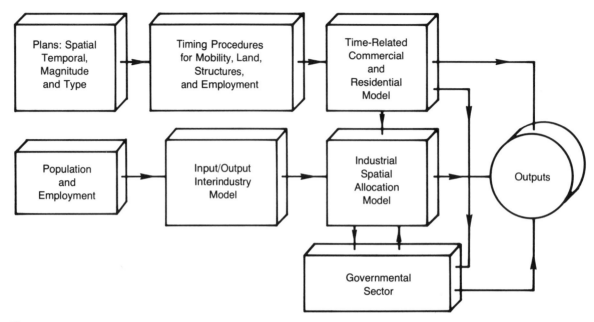

Figure 3-4

emphasize the measurement and prediction aspects of urban-renewal decisions over time. The four parts necessary to accomplish this are:

1. Design alternative plans for urban development.
2. Evolve alternative urban-renewal programs which will accomplish the plans of part 1.
3. Compute the several effects for each of the alternative plans developed.
4. Choose the preferred plan of (1) based on the information gathered in (2) and (3).

All four parts of this process will involve computerized models. We have discussed simulation models in a general way in Chapter 2, and now more details are necessary, especially with respect to the information requirements of the models.

The simulation model of interest is based on an earlier model developed by Lowry (*11*). We referred to this model previously as an example of a combination allocation and derivation model. Lowry's Model of Metropolis deduces the distribution of residential and residence serving activities. The submodels of the overall simulation model are depicted in Figure 3-4 in the form of a block diagram showing the flows of information. Outputs of the model format in Figure 3-4 include

A. Annual census information
1. Population by type.
2. Employment.
3. Land use.
4. Social indices.
5. Project direct impacts.
6. Housing types.
7. Surplus and deficit industrial, commercial, and residential markets.
8. Income.
9. New firms with growth or decline.
10. Blight indices.
B. Annual or biannual city information
1. Revenues by type.
2. Expenditures by type.
3. Fiscal reconciliation alternatives.
4. Project-oriented information priorities.

There are three major assumptions associated with the simulation model of Figure 3-4. The discussion of these assumptions will shed more light on the input–output details of the blocks in the simulation flowchart. The *first assumption* is that employment opportunities are directly or indirectly responsible for most development decisions. The input–output interindustry model of the figure performs the function of projecting employment for the region, county, and city for various employment categories; then, based on workforce participation and patterns of travel to work, the total population of the city is estimated. These results of the input–output interindustry model are compared to the block designated "Population and Employment," which produces a population estimate according to age, sex, and race based on small area projections.

The *second assumption* associated with Figure 3-4 is that there is an employment base which forms and exists because of several locational criteria, including existing employment grouping, land-use policies, transportation access, and land-assessment patterns. These locational criteria can be changed until a future block of employees is distributed for the first time period. If the future "site-oriented" employment cannot be accommodated by the manipulation of location criteria, the assumption is that the employment leaves the city. The process just described is performed by the block designated "Industrial Spatial Allocation Model."

The remaining employment, which is residential (not site-oriented) is dealt with by the *third assumption*. The effect of employment and the location of households is calculated based on the facts that households cluster at fixed distances from employment, and, in turn, commercial services group in a known proximity to residential areas. Other factors considered in the location of households are racial, occupational, and economic characteristics and type of housing. The process we have been describing is represented by the block "Time-Related Commercial and Residential Allocation Model." Its function is to transform new employment into new households, which produce new service employment, and so on cyclically until stability in the process is reached. The model is constrained by existing residential and commercial clusters and land-use policies.

The discussion of the model thus far must have indicated to the reader the extensive need for data collection, organization, and analysis. As we review the data needs, contrast them with the growth-allocation model of the previous section.

The heart of that model was a theoretical statement, but in our current model we have no such inclusive base. Instead we have reasonable and workable assumptions and an extensive data base. The comparison of the urban models of this and the previous section is a graphic case in point of the difference between a simulation and an analytical model.

The data utilized for the Pittsburgh simulation model included

1. In-depth surveys of Pittsburgh.
2. Detailed interviews with a large number of respondents.
3. Detailed U.S. Census data for 1960.
4. Extensive data from the Department of City Planning on its property inventory.
5. A variety of miscellaneous data on industry, the labor force, government, and so on.

The statistical tool used to analyze much of the data was factor analysis, and, although not discussed in Chapter 2, factor analysis is an extremely useful tool for the practitioner of systems modeling, particularly simulation modeling. Factor analysis is a technique used frequently by social scientists to simplify large arrays of variables (*16*). Each variable is correlated with every variable of interest, and in the resulting matrix array, clusters of the variables may suggest an underlying factor common to all of them. Often data are too extensive to allow judgmental inferences; however, with the statistical technique of correlation analysis, which produces a table of correlated factors, followed by factor analysis, which gives a simplified and reduced version of intercorrelations, critical and related variables can be detected. What we have decribed is a data-manipulation technique, and the grouping of variables according to factor-analytic procedures does not ensure the validity of an inventory of variables collected in that way. Separate demonstrations of validity, such as empirical evidence, are always necessary.

In the Pittsburgh model, factor analysis had two purposes. The first was to identify the critical factors from among the many factors contained in the census data, land-use data, and the various surveys. The second purpose was to identify categories of census tracts in a way that the various tracts were characterized according to the intensity of the variables included in each tract. In factor-analyzing households, housing, commercial and industrial activity, and combinations of demographic–economic–social–physical data, over 600 separate variables were considered.

In addition, regression-analysis techniques were applied to much of the data. The purpose of regression analysis is to infer a descriptive relationship between observed data, particularly between a presumed dependent variable and one or more of a set of independent variables. This is done by fitting an equation to the observed data. Five of the most important pieces of data that were regression analyzed are

1. Household-housing types: The dependent variables were taken as the types of housing, and the value, rental, age, and size of units involved; the independent variables were household types based on income, race, age, and size.
2. Household mobility factors: The dependent variables are the rate at which household

types move across the city; the independent variables are various household types as in item 1.

3. Commercial activity analysis: The component factors measured from the commercial survey are the dependent variables; the coefficients of access to each tract and the population distribution by household type in each tract are the independent variables.
4. Industrial spatial analysis: The employment of each industrial component in each census tract was the dependent variable, and the independent variables are relevant industrial attributes such as spatial factors, condition, assessed value, and accessibility factors.
5. Municipal expenditure and revenue analysis: The levels of various types of revenues and expenditures serve as dependent variables, and these are related to population, income, employment, housing, and commercial and industrial activity.

The model is composed of the extensive data reviewed and the various data-processing procedures discussed. There are over 30 computer routines, which perform the following functions:

1. Data manipulation and management.
2. Statistical analysis of data for the locational aspects of the model.
3. Behavior simulation of locating groups.
4. Monitor environmental conditions, and communicate between locational models.
5. Summarize, interpret, and display the results of model runs for the use of model builders and decision makers.

Clearly, the Pittsburgh simulation model as we have described it is large, extensive, and appears cumbersome to operate. These observations are valid, and the model can be used to test only a limited number of policy alternatives. The refinements of a simulation model, which, in effect, continue throughout the life and use of the model, include better ways to deal with all aspects of data, more understanding of the details of the particular phenomenon being modeled, improved computer routines, and more attention to validation of the model.

The Pittsburgh urban-renewal simulation model allows the prediction of residential locational choice on the basis of job location, and predicts the location of commercial activity on the basis of residential location. The actual construction of the model has given the participants a learning interdisciplinary experience, and provided the urban development literature with a useful contribution.

Evaluation of Urban Performance (17)

One paramount intention of all urban modeling is to determine the answer to the following question with respect to the urban setting: How are we doing? The answer to this question is found first by understanding, as well as possible, the urban process in question, and second by evaluating what is happening by using some type of performance measures. Needless to say, evaluation of even a well-understood complex process is very difficult. An evaluation will probably be done using a set of criteria because of the interactive nature of the urban system. For example, improved travel mobility for an urban area is likely to be accompanied by negative impacts, such as neighborhood disruption and increased pollution. In this, the last major section of Chapter 3, we want to consider methods of measuring urban performance.

Urban planning procedures, including some we have discussed, often utilize computer program packages to check and organize urban data, forecast travel volumes, and analyze transportation networks, among other things. These programs provide little or no information on the social impacts of the related urban programs. This is a recognized limitation in the current generation of urban development models. The evaluation methodology of this section is aimed at dealing with that limitation.

A useful evaluation methodology would permit the comparison of alternative urban policies in a total way, not just with respect to direct urban service adjustments. The methodology will be developed in three steps:

1. The major attributes desired by the residents of an urban area will be identified and quantified in macroscopic terms.
2. Using these desirable urban attributes, the overall performance of the urban area is described in terms of the availability of these attributes to different groups of its residents.
3. Alternative planning policies and programs can now be evaluated in terms of changes that effect the availability of the urban attributes.

Clearly, the position of this methodology is to determine what the urban dwellers want, then find out how well those desires are being met as a measure of urban performance. An alternative

conceptual direction would be to find out what the people should have according to a value structure other than their own. Similarly, an idealized urban system could be designed according to some acceptable criteria, then the existing urban system compared to it to evaluate a measure of urban performance.

Urban attributes are divided into two groups: metropolitan attractions, and local or neighborhood characteristics. Metropolitan attributes include such things as employment, shopping, and recreation. The attribute description must be in terms of quantity, quality, and accessibility. Neighborhood attributes list items essentially independent of the surrounding urban structure, such as quality of local streets, housing, schools, shopping, and fire, police, and other local services. Methods for describing the availability of these attributes will give values as an availability measure for various areas and resident groups in the city, and the way in which the values are changing will be construed to be a measure of urban performance.

Beginning with the first step in the methodology development, we want to identify and quantify urban attributes. Urban attributes will be termed "metropolitan opportunities" and will be identified in terms of residents who can take advantage of them because the residents possess the requisite funds, skills, or background to make the attributes desirable to them. The second component of metropolitan opportunity is urban friction, which is the measure of the difficulty of access of the attractions. We shall measure metropolitan opportunity in the following way. Let the urban area be divided into a number of geographical neighborhoods, designated $i = 1, 2, 3, \ldots$. Divide the residents of the urban area into groups according to income, race, years of education, or other criteria and designate them as $p = 1, 2, 3,$ .... Then

A_{jp} = measure of the metropolitan attractions appropriate to resident group p in zone j,

Q_{tjp} = quantity of metropolitan attractions of type t in zone j appropriate to resident group p,

where $t = 1, 2, 3, \ldots$ and $t = w$ when the urban attraction is work. The quantification of Q_{tjp} might, for example, be square feet of shopping space if the attraction being considered is shopping. A_{jp} and Q_{tjp} can be related in the following functional relationship:

$$A_{jp} = f(Q_{1jp}, Q_{2jp}, \ldots).$$

One other ingredient is necessary to formulate A_{jp}. Not all attractions are of equal importance to urban residents, and so a weighting term, B_{tp}, will be used to account for the importance of an urban attraction t to resident group p. Assuming the functional relationship above to be a linear summation, we can write the total attraction as

$$A_{jp} = \sum_{t} \frac{B_{tp} Q_{tjp}}{Q_{t\text{-}p}},$$

where $A_{t\text{-}p} = \sum /j\, Q_{tjp}$. The term $B_{tp}/Q_{t\text{-}p}$ expresses the importance to resident group p of a unit of attraction t in the urban area. The most common attraction is employment, so that the importance of one unit of employment attraction for resident group p is denoted

$$\frac{B_{wp}}{Q_{w\text{-}p}}.$$

In order to evaluate metropolitan attractive-ness, A_{jp}, data must be available to determine Q_{tjp} and B_{tp} values. It is assumed that values of B_{tp} can be obtained directly from primary data sources such as home interview surveys. Evaluation of Q_{tjp} is based on other data usually available for most large U.S. cities. Typical information required includes total jobs appropriate to a particular resident group, fraction of attractions appropriate to a resident group occurring in land use, and units of activity of a particular kind of land use in a given zone. Using this kind of data, and other kinds derivable from it by intermediate steps, Brown and Kirby are able to calculate values of A_{jp}. In an example given in their paper they compute the attractiveness in job equivalents of three urban zones from the viewpoint of one group of residents who are subdivided according to income level. Results allowed the formulation of statements such as: the total attraction provided by zone j is greatest for the middle-income group, with less attraction for the high- and low-income groups. Specific data were available to substantiate this.

The second dimension, in addition to attraction, which must be determined for metropolitan opportunity is urban impedance. That is, we must quantify the difficulties encountered by residents in traveling between zones to take advantage of these attractions. Such a measure is intended to reflect both the transportation services available and the manner in which the residents take advantage of them. Urban impedance will be designated d_{ijp}, denoting a measure of the urban travel impedance encountered by residents of group q in traveling from zone i to zone j. The calculation of d_{ijp} is made from the following definition:

$$\frac{1}{d_{ijp}} = \sum_v \frac{P_{pv}}{d_{ijpv}},$$

where d_{ijpv} = measure of the urban travel impedance experienced by residents of
group p in traveling from zone i to
zone j by mode v,

P_{pv} = factor representing the preference
of resident group p for mode v.

Other forms of d_{ijp} exist which account for the
tradeoff between travel time and travel cost by
each mode for each resident group.

The two dimensions of metropolitan opportunity, attractiveness and urban impedance, have
been defined in a way so that quantification is
feasible. Now we must combine these two values
to give a composite measure for metropolitan
opportunity. We define T_{ip} as a measure of the
total metropolitan opportunity for resident group p
in zone i as

$$T_{ip} = \sum_j \frac{A_{jp}}{d_{ijp}}.$$

The term T_{ip} has been defined in such a way that
it reflects all the attractions appropriate to resident group p in the urban area, adjusted for the
travel difficulties encountered by the resident
group living in zone i and traveling to the attractions in each zone j. Changes in any aspect of
attractiveness or the impedance of the urban environment will be reflected in the T_{ip} values.
These variation in T_{ip} values will serve as a
measure of the effects of such changes in the urban environment.

The second major aspect of urban attributes

is neighborhood quality, and its measure will be
discussed at this point. The definition of attractiveness did not include any attributes of the
immediate neighborhood such as local street appearance, quality of nearby schools, and effectiveness of police, fire, and sanitary services.
In order to quantify neighborhood quality we assume that there are measurable qualities, q_i,
which characterize the neighborhood in the
sense we have just discussed. We also assume
that the q_i values will be weighted to account for
the effect on a zone of the characteristics of surrounding zones. Within these considerations a
useful of N_i, the neighborhood quality is

$$N_i = \sum_j \frac{q_j R_j / d^2_{ij}}{\sum_j R_j / d^2_{ij}},$$

where R_j = total residents in zone j,
d_{ij} = measure of the proximity of zone
j to zone i.

Using the ideas of metropolitan opportunity
and neighborhood quality, the various functions
may be computed for the zones in an urban area.
The result is a very special socioeconomic view
of the city based on demographic characteristics,
the distribution of land use activity, and the transportation system. Actual or proposed changes
may be evaluated using the defined variables.
One generalization that Brown and Kirby noted
was that changes in the transportation system
or in location of metropolitan attractions will affect
metropolitan opportunity, whereas changes in
housing and urban services will affect neighborhood quality.

This section has reviewed an evaluation framework for urban performance which assumes that a primary function of a city area is to make urban attributes available to its residents. The details associated with attribute identification represent one possible set of quantitative descriptors which could fit in the methodological framework. Whether a particular set of attributes are appropriate for a specific urban area and whether the values for the measures represent good, fair, or poor performance must be argued by those concerned with the present and future life of the area. If sufficient urban data are available, answers to questions of the following type are possible. Are specific neighborhoods improving or declining in quality? Would new transportation modes or land uses maintain or improve metropolitan opportunity value? What resident age groups are affected by changes?

The evaluation of urban performance by measuring changes naturally focuses attention on the possible change agents. Urban plans and policies and land locational decisions are potential change agents. The implementation of this evaluation framework was based on the fact that many urban decisions are strongly influenced by factors that correspond conceptually to the measure of metropolitan opportunity and neighborhood quality in the urban zones.

Aftermath

The Lowry model (*17*), developed in 1962–1963 and reported in 1964, represents a first-generation result in urban modeling. Now, more than a decade later, it is appropriate to take stock of the stream of work which came from the model "Metropolis." The formation of the model amounted to a breakthrough of a particularly intractable conceptual barrier (*18*). Although there is record of only a few domestic users, the applications abroad have been extensive. This modest explosion of successors to the Lowry model was prompted by three attributes of the model. There was the distinct promise of meaningful operationality; there was a causal structure, which was appealing simple; there was clear opportunity to expand and modify the framework. Goldner in reference (*18*) points out, that although at times discouragingly small, there is an expansion of the use of advanced analytic techniques in public decision making.

A list of the domestic descendants of the Lowry model include: Time-Oriented Metropolitan Model, 1964; Bay Area Simulation Study, 1965; Garin–Rogers Contributions, 1966; The Cornell Land-Use Game, 1966; A Dynamic Model of Urban Structure, 1968; and Projective Land-Use Model, 1968. Reference (*18*) reviews each of these models.

In England, Lowry-type offshoots cover at least six subregions, and deal with such problems as the impact of new towns, the location of major airports, and the planning of urban structure within the framework of a prospective reorganization of local government. English subregions correspond to medium-sized metropolitan areas such as Albany–Schenectady–Troy.

Although many of the models that followed the first-generation results did not reach the operational stage, the studies that were done confirm the promise of modeling and the systems approach for the planning and operating functions of urban areas.

References

(*1*) H. B. Wolfe and M. L. Ernst, Simulation Models and Urban Planning, in *Operations Research for Public Systems,* P. M. Morse, ed., The MIT Press, Cambridge, Mass., 1967.

(*2*) B. M. Gross, What Are Your Organization's Objectives: A General Systems Approach to Planning, *Human Relations,* 195–217 (1965).

(*3*) H. W. Maier, The Troubled City, Chap. 6 in *The Challenge to Systems Analysis,* G. J. Kelleher, ed., John Wiley & Sons, Inc., New York, 1970.

(*4*) A. G. Wilson, Models in Urban Planning: A Synoptic Review of Recent Literature, *Urban Studies,* **5**(3), 249–276 (1968).

(*5*) I. S. Lowry, A Short Course in Model Design, *Am. Inst. Planners J.,* **31**(2), 160 (1965).

(*6*) D. B. Suits, Forecasting and Analysis with an Econometric Model, *Am. Econ. Rev.,* 104–132 (1962).

(*7*) M. D. Kilbridge, R. P. O'Block, and P. V. Teplitz, A Conceptual Framework for Urban Planning Models, *Management Sci.,* **15**(6), 246–266 (1969).

(*8*) N. A. Irwin, Review of Existing Land-Use Forecasting Techniques, *Highway Rev. Board Record,* No. 88, 187–189.

(*9*) N. A. Irwin, Review of Existing Land-Use Forecasting Techniques, *Highway Rev. Board Record,* No. 88, 184–187.

(*10*) T. G. Donnelly, F. S. Chapin, Jr., and S. F. Weiss, A Probabilistic Model for Residential Growth, Institute for Research in Social Science, Chapel Hill, N.C., May 1964.

(*11*) I. S. Lowry, A Model of Metropolis, Rept. RM-4035-RC, Rand Corporation, Santa Monica, Calif., Aug. 1964.

(*12*) K. J. Schlager, A Land Use Plan Design Model, *Am. Inst. Planners J.,* **31**(2) (1965).

(*13*) C. Alexander, *Notes on the Synthesis of Form,* Harvard University Press, Cambridge, Mass., 1964.

(*14*) D. M. Hill, A Growth Allocation Model for the Boston Region, *Am. Inst. Planners J.,* **31**(2) (1965).

(*15*) W. A. Steger, The Pittsburgh Urban Renewal Simulation Model, *Am. Inst. Planners J.,* **31**(2) (1965).

(*16*) B. Kleinmintz, *Personality Measurement,* Dorsey Press, Inc., Homewood, Ill., 1967.

(*17*) A. Brown and R. F. Kirby, Measuring Urban Performance, presented at the 40th National Meeting of the Operations Research Society of America, Oct. 27–29, 1971.

(*18*) W. Goldner, The Lowry Model Heritage, *Am. Inst. Planners J.* (Mar. 1971).

CHAPTER 4

Transportation Systems

Literature Resources

The literature of transportation systems is voluminous, possibly the largest segment of the urban development literature. The spectrum of transportation models presented ranges from simple speculation to complex simulation models. The literature deals primarily with two topics. One type of study emphasizes the structure and dynamics of discrete elements of a transit network and this kind of work is usually done by transportation engineers. The second kind of study looks at the transit problem as it is related to other urban problems, and this sort of effort requires the participation of many skills and disciplines.

Current emphasis of research on problems in the area of transportation systems is in four categories:

1. Mass transportation systems designed to carry large numbers of people as economically and as efficiently as possible on land. Large amounts of federal funds are being invested in this category.
2. The private-vehicle transportation system, designed to move individuals in and through less densely populated areas. Currently considerable effort is being invested in the use of computers to control the flow of vehicular traffic.
3. The highway system, whose function it is to carry people in and through sparsely populated areas. Past work has been dominated by civil engineers, who defined the problem as designing the most economical, safest, and beautiful highway. An alternative per-

spective recognizes that a highway system allows the interaction of people, machines, and environment in an extremely complex way, and studies from this viewpoint are being done.

4. The air transportation problem. The issues of this category are concentrated on the control of air traffic in flight and entering and leaving the terminal under all-weather conditions. A serious, unsolved aspect of air traffic systems is the inability to remove weather as a factor in the flow of traffic. No operational all-weather landing system exists capable of being used by commercial air carriers.

Nearly every American city of over 1 million population has financed transit studies. Several good simulation models exist for cities such as Chicago and Pittsburgh. The State of California as part of the social-problem-solving tie to the aerospace industry, which we reviewed in Chapter 1, contracted for a state transportation study. Five volumes appeared in 1965 and presented a plan for a large simulation model to cover the transportation needs of the entire state. The project was never begun, because funds were not available.

Reviewing the transportation literature of past years, it appears an inescapable fact that any form of transportation that offers the lowest door-to-door travel time will always drive out lower-speed competing modes unless the economics of the higher-speed system are grossly unfavorable. Implicit in this observation is the tie to our national life styles and cultural norms. Problem solving in the transportation-system area is characterized by the need to obtain an intimate assessment of human values regarding such things as travel time, safety, beauty, and so on. The evaluation of these "human factors" must include economic considerations, and efforts to quantify these values have not been very successful.

The user of the transportation-systems literature will face a number of difficulties. Most modelers will be constrained by limitations on time, manpower, money, and the need for a model sufficiently accurate to provide reasonable predictions. The literature is likely to provide accurate models but not ones comprehensive enough for the modeler's needs. Other models available are accurate and comprehensive but require excessive data and manpower requirements. Often system studies have been aimed at particular local or regional transportation problems with the results in the form of specific recommendations with little or no attempts at generality. Any transportation relationships formulated for an area other than the area of current interest must be tested carefully and validated.

Recent contributions to the literature have considered the evaluation of several alternative transportation-modal mixes. Two aspects of this kind of research have been (1) the attempt to understand the functioning of people in high-density situations and the permissible transportation under these conditions, and (2) the design of an optimal transportation network in terms of maximum efficiency and minimum contribution to other areas, such as public health. The second task can be done only with the results from the first task.

This opening section on the extensive literature

of urban transportation is presented to lend substance, in part, to a growing recognition that many of the ills of our cities stem from the problem of transportation within the metropolis. Many qualified observers speculate that the transportation dimension of the urban structure is the single most important and critical variable which must be understood.

Urban Transportation Systems

Problems are inseparably associated with urban transportation systems. Records of Rome dating back to the first century A.D. included vehicular restrictions to attempt to deal with congestion. Urban traffic difficulties did not come to the Western world until the nineteenth century, when industrialization was well under way. The contemporary complaints commonly heard about modern systems of urban transportation are congestion, the overloading of routes and facilities, the overlong trips, the irregularity and inconvenience of those services that are publicly provided, and the difficulty of parking private vehicles at desired destinations. Problems of this ilk are a consequence of the enormous size of urban areas, the organization of land use, the rhythm of urban activities, the balance of public services with private rights of access and movement, and the preferences and tastes of people with regard to mode of travel, route, comfort, and cost. Transportation problems are an integral part of urban space difficulties, of investments, and of the general effort to meet the needs and aspirations of the city dweller. In spite of this, traffic congestion is the primary image of transportation problems to the average person.

Industrialization in an open society where there is free choice of work and free choice of the journey to work supplied the origins of the modern traffic problem. The older structure of groups of homes clustered close to the places of work has given way to a desire to improve living standards. The dispersion of homes to seek better living environments has contributed almost as much to urban traffic problems as the actual growth of the cities (*1*). In all prosperous countries where the standard of living has been increasing, the average citizen has had more money to invest in transportation and has come to expect greater comfort and convenience. With this exercise of individual opportunity has come a host of difficulties, including urban transportation problems. For this reason any serious considerations of major changes in urban transportation must confront the entire issue of the quality of urban life and the life styles of its dwellers, whether they be residents or workers.

After this previous discussion, which, to some extent, emphasized traffic congestion, it is proper to recall that the purpose of an urban transportation system is to move people and goods from point to point. We must think, therefore, of the terminations as well as the pathways. The city planner has to view the transportation parts of city layout as having two aspects. The first is to determine the tasks and requirements of the transportation system, and the second is to find ways of achieving those objectives within social and economic constraints. Land-use policies and their effect on transportation needs is a natural place to look. Many land-use studies of American cities have been done to determine the effect of land-use policies on the generation of different or variable traffic flows. Urban land-use studies of this kind began in the 1950s and had the effect of shifting the focus from the study of the traffic flows to a study of the land uses that caused the flows. The emphasis on the functional relationship between the city and transportation has come to dominate large urban transportation studies. The merging of data in these studies from extensive surveys on both traffic and land use has made four major contributions:

1. Transportation is viewed as a comprehensive set of interrelated activities.
2. Land use, demographic and social characteristics, and consumer choices are recognized as major determinants for transportation requirements.
3. An awareness was developed for the effect of the transportation system itself in influencing the development of urban areas.
4. An acceptance of the fact that metropolitan scale planning, development, and control of transportation systems is necessary in an environment where the citizenry move freely across the boundaries of local government.

The study of the functional relationships between users and facilities is aimed at solving problems associated with obtaining efficient movement through the transportation network, and learning how to promote new urban activities. Even if the functional relationships were well known, the understanding would be of no value unless there were clear policies on such issues as: should the cities be centralized or dispersed, and should multimode transportation systems with particular constraints (such as the restriction of the automobile) be implemented?

The influence of new transportation systems on the formation of the city is well known. Dispersion of cities first occurred because of subways and street railways, and it now continues at an incredible rate because of the automobile. If urban growth has a "natural" dispersion character, this is one matter. However, if supporting the use of the automobile creates the dispersion with painful side effects, a different set of issues must be resolved. City planners are divided on this point.

Projections from every major transportation study indicate that the use of the private automobile will increase. With this increase must come a heavier volume of traffic, construction of new highways, and further dispersion of the metropolitan population. In spite of this, many cities are involved in new investments or reinvestments in public mass transportation. These efforts have come at a time when the fortunes of the public transportation industry are at a low ebb. Because public transportation is becoming more important in urban transportation schemes, it will be useful to review some aspects of mass-transit systems.

As our cities grew during recent decades three difficulties of urban transit emerged. These are the collection problem, the delivery problem, and the peak-load problem. Urban sprawl, low density distribution of residential populations, and widespread car ownership has caused the collection problem. Ideally there is mutual support between an urban transit system and the nearby people it serves. The system is there because of the people, and the people are there because of the available service. Efficient collection of dispersed people is simply not possible. Low collection densities maximize the operating disadvantages of the fixed-rail transportation system—low efficiency at low operating speed, high cost of braking and acceleration, problems of scheduling, and the minimum profitable payload required by fixed costs.

The problem of delivery exists due to the distribution of places of work and general activities in the downtown area. The shortage of land in the central area has caused the migration of industries to the outskirts of cities. The distance from well-dispersed delivery points to central city activities has become prohibitively long.

The peak problem is the well-known difficulty associated with the fact that journeys are not distributed in time but are concentrated at certain preferred hours. Transit companies have reported that 80 percent of their volume of travel is concentrated in 20 hours of the week. If the peak-load requirements are successfully met, the system will be underutilized at other times.

Beyond the three problems of urban transit just discussed are a whole array of negative factors associated with public transportation. Included are changes in consumer tastes and expectations with respect to comfort, convenience, privacy, storage capacity, guaranteed seating, and a number of other social intangibles. Most of these requirements favor the use of the private automobile.

If we view the quandary of public versus private transportation from an economic viewpoint, it appears that the automobile commuters are making a rational choice. In spite of the promised lower cost of mass transit, economic studies concluded that the car is economically competitive with other available modes of travel to work. Assump-

tions included a relatively high rate for the driver's or passenger's time.

With the continuing growth of national income, the expanding use of the automobile is likely to continue. Cities are using three strategies to deal with this condition. One effort is to provide all-out accommodation of the automobile, and the second is to ban the automobile from center city and replace it with rail mass transit. Other strategies lie between these two extreme positions. Any of these possibilities appears to be affected by the adage: additional accommodation creates additional traffic. The real issue is not trains versus cars but how to balance the two. Government support favors the strengthening of rail systems while consumers continue to vote for the use of the automobile. Within this setting we will review what has been and is being done.

Traffic System Analysis

It is our intention in this section to review the methodology available for traffic system analysis. These tools are used by both engineers and planners and reference (2) is an example of a source of the methodology from the literature.

Both the traffic engineer and the planner make decisions that have an impact on the everyday life of large populations. By practical necessity they must be involved with economic, social, political, governmental, and physical features of their environment. For these reasons *rational* decisions made within a *formalized* framework for analysis and synthesis are required. The framework utilized is the systems approach, and alternative designs and decisions evolved in this way will have all the features we have discussed previously and, in addition, will be capable of evaluation. The tie of the traffic engineering design process to the systems approach is best illustrated by the following steps of the design process (2):

1. Selection of goals and objectives.
2. Determination of elements or variables.
3. Formulation of model.
4. Search for, and specification of, alternative designs.
5. Measurement and evaluation of consequences of alternative designs.

And, of course, associated with the five steps are the requirements for data collection and analysis.

Within these five steps a number of processes and techniques are used in order to achieve the final design or decision.

1. Statistical data must be taken and analyzed.
2. Linear regression analysis must be done to establish relationships between variables.
3. Traffic flow and volume must be predicted for varying demand conditions.
4. Economic analytic techniques must be applied to alternative designs.
5. Highway and congestion cost and pricing principles must be understood.
6. Theoretical techniques for describing traffic flow must be applied.
7. Performance characteristics of highways must be measured or designed.
8. Signalized and nonsignalized intersections must be analyzed.
9. Simulation models must be designed.

This set of nine techniques specifies the kinds of skills and information that must be available to accomplish the five steps associated with the systems approach.

Returning now to the five steps of the design process, we will isolate and discuss those points unique to the solution of a traffic problem.

In problem definition, goals and objectives must be expressed in explicit terms. In order to determine how well alternative designs compete in satisfying goals, *measures of effectiveness* are utilized. Common measures of effectiveness include maximum flow or volume, minimum average vehicle delay or travel time, minimum stops per hour, minimum accident rate, minimum backup, minimum annual cost, maximum rate of return, and maximum net present value. Effectiveness in terms of safety of human life or aesthetics must be included but are much more difficult to quantify.

The models used in traffic studies are mathematical generally, and digital-computer simulations are common.

Within a variety of constraints in the form of "standards" and "norms" the search for alternative designs takes place. The evaluation of the consequences of alternative designs includes consideration of determining the costs and benefits of systems of differing physical makeup, and of varying levels and kinds of service. The intangible aspects of social dislocation, aesthetics, and public safety must be included even if oversimplified procedures are developed. The values placed on social objectives such as these appear to vary from time to time and from decision to decision.

Often there are multiple and conflicting design objectives. The design of a high-speed expressway to reduce the travel time between two points is likely to incur an excessive commitment of capital funds and other resources. Whenever the conflicts involve social objectives, the involvement of a team of professionals, including sociologists, psychologists, and economists, is essential. Here the social value issue arises again. Should the collective value structure of the community be applied to aspects of traffic system designs or should some other value structure be invoked?

The variables in the design and the model will depend on the measure of effectiveness to be used. The general categories of effectiveness are safety, speed, comfort, convenience, and economy. Therefore, the key variables to be used in describing traffic movement are

1. Composition: What are the nature and characteristics of the moving traffic?
2. Volume: How much traffic is moving?
3. Origin and destination: Movement patterns.
4. Quality: How well is traffic moving?
5. Cost: What is the expense of moving the traffic?

Each of these five variable sources requires some amplification. The composition of traffic includes vehicles for the movement of people and vehicles for the movement of goods. The nature of the composition of traffic can affect facility design in two ways: human factors and vehicle factors. Human factors requires consideration of such aspects as vision, strength, reaction capability, motivation, temperament, attention, and so on. Examples of vehicle factors are all the vehicle physical dimensions, such as length, width, and weight, and braking power, turning radius, acceleration, and so on.

Data on the volume of traffic range from simple rates of flow and their duration, to peak-volume phenomena, directional movements, and descriptions of purposes for which the trips are being made.

Movement patterns have almost as much effect on facility design as does volume. Movement patterns influence the necessity of turning lanes, intersection turning movements, distinction between through and local traffic, and aid in the specification of ramp spacing.

Quality of traffic movement is difficult to define precisely. Factors that merit consideration are speed, convenience, comfort, safety, and privacy. Quality involves the motivation of travelers and their reaction to level of service. No standardized quantification schemes exist for quality factors. In a given study relevant quality factors will be identified and dealt with by some means unique to the phenomenon. Examples will be given later.

Two categories of the cost of traffic movement are usually specified. The first are the costs from the viewpoint of the traveler, and the second has to do with a complete accounting of all the costs that must be incurred by some person or agency to permit movement of traffic. The categories emphasize the need for an appropriate frame of reference when doing any cost benefit analysis.

The comments on traffic system analysis thus far have been largely philosophical. Now we want to examine some of the methodology used to implement the ideas. We want to develop symbolic models in the form of relationships to provide a starting point for analyzing the characteristics of the driver–vehicle–highway system and the volume, density, and speed variables. With relationships between variables known, the characteristics of traffic streams can be determined and the consequences of alternative designs predicted. The fundamental relationship will relate volume, density, and space mean speed, and these variables are defined as follows:

Volume: the number of vehicles passing a point in a unit of time.

Density: the number of vehicles traveling over a unit length of highway.

Space mean speed: the arithmetic mean of the speeds of vehicles occupying a given length of roadway at a given instant.

If we consider a short length of roadway designated x feet, for time intervals T seconds and observe n vehicles crossing a line M located in x, then we can write

$$\text{volume rate} = V = \frac{n}{T} \text{ cars per hour.}$$

The density D, can be expressed as

$$D = \frac{\text{av. no. vehicles traveling over } x}{x},$$

and the units of D will be vehicles per foot. The average number of vehicles traveling over x can be calculated by summing the times required for each vehicle to pass over x, and then dividing by the time interval T, or

$$\text{av. no. vehicles over } x = \frac{\sum_{i=1}^{n} t_i}{T},$$

where t_i is the time for the ith vehicle to move the distance x. Now we can write the density as

$$D = \frac{\sum_{i=1}^{n} (t_i/T)}{x} \text{ vehicles per mile.}$$

Space mean speed, U_s, is expressed as the quotient of volume and density, or

$$U_s = \frac{x}{1/n \sum_{i=1}^{n} t_i} \text{ miles per hour,}$$

or, in a simpler form,

$$V = DU_s \text{ cars per hour.}$$

This equation is very general and potentially useful; however, each variable is dependent upon a myriad of physical and psychological parameters that are influenced by the particular sample of drivers, vehicle characteristics, weather, and road conditions, among many others. From this fundamental symbolic model come a variety of other equations, which provide a theoretical foundation and supply a framework for empirical studies.

As a second example of traffic systems methodology we will look at a method of forecasting the demand for traffic facilities. The tool to be used is linear programming, which was reviewed briefly in Chapter 2. Clearly, in a random process such as traffic, we expect the volume to vary considerably during a given period. Therefore, the requirement for suitable traffic facilities may appropriately be called a *time-dependent demand function*. One group obviously concerned about this is any toll authority, and we will review the systems methodology by using it to determine the uniform toll rate necessary to provide sufficient revenues to just cover bond payments or capital plus interest costs for the facility. We won't consider maintenance and operating expenses. The variables of this problem will be related mathematically, and three aspects of the problem must be included:

1. The hour-to-hour traffic demand functions in which the quantity demanded for a given hour is a function of the toll plus other *traveler-perceived* payments of time, effort, and expense
2. The user price–volume relationships
3. The roadway toll function or toll as a function of traffic flow and fixed facility costs.

Forming relationships we have (2)

$$q_i = f_1(p_i),$$

$$p_i = f_2(q_i, t),$$

$$t = f_3(F_x, q_1, q_2, \ldots, q_{24}),$$

where
- i = hour and $i = 1, \ldots, 24$,
- q_i = quantity demanded during the ith hour,
- p_i = toll plus other perceived payments per trip during the ith hour,
- t = uniform toll per trip charged during all hours,
- F_x = fixed-cost portion of facility x.

The mathematical relationships we have just defined can be combined into a linear model for demand, price–volume, and toll function in the following way:

$$q_i = a_i - b_i p_i,$$

$$p_i \geq c_j + d_j q_i + t, \qquad j = 1, \ldots, m,$$

$$t \geq f_e - g_e \sum_{i=1}^{24} q_i, \qquad e = 1, \ldots, n,$$

where
- a_i and b_i = demand coefficients during the ith hour,
- c_j and d_j = cost coefficients for the jth user price–volume,
- f_e and g_e = cost coefficients for the eth fixed-facility cost.

The model is somewhat imposing, so let's review the general process by which it was obtained. The problem was studied and data were taken until understanding of the process started to emerge. Variables were identified and simple functional equations were formed to relate appropriate variables, then the relationships were combined to form expressions that were symbolic mathematical models of precise verbal statements which could be made about the process in question. The equations are developed in a linear programming formulation, and we want to apply them to a specific toll facility. To do this we must determine the appropriate demand and price–volume functions for the particular case, the facility costs, and adjust the data so that they can be used in the linear programming technique.

Traffic flow is a process we cannot describe with certainty. Therefore, probabilistic models based on extensive data taking and analysis exist in many forms in the literature for nearly every aspect of traffic flow. One well-known class of traffic models is the queuing model, which provides a mathematical description of waiting lines of traffic and their delays. The queuing models are formed to answer the following kinds of questions:

1. What is the probability that an automobile arriving at a specified point will be delayed or be faced with a queue?
2. How much waiting time can be expected?
3. What is the rate of dissipation or establishment of a queue?

A sample queuing model is given next for $p(n)$, the probability of having exactly n vehicles in the traffic system:

$$p(n) = (\frac{\lambda}{\mu})^{n} (1 - \frac{\lambda}{\mu}),$$

where λ = average number of vehicle arrivals per unit of time,

μ = average number of vehicles discharged per unit of time.

In the discussion thus far on traffic analysis, the viewpoint has been limited to micromodeling. That is, the concern has been to understand the details of the actual traffic flow rather than the integration of the traffic flow into a complete transportation system. This is the emphasis of traffic engineering where facilities are designed to meet the demands of the actual traffic. In the next section we will look at the total system. The large number of traffic-flow models are used in traffic simulations as an aid to facility design. The block diagram in Figure 4-1 should be a familiar format, and in this case the systems approach is applied to traffic simulation.

Transportation Systems

The basic issue to be considered in this section is ways of examining total transportation systems to assess their worth. In other words, we want to do *evaluation*. Evaluation is often a natural part of the application of systems approach, and there as elsewhere, evaluation is frequently controversial. The general process of evaluation may be characterized in the following steps:

1. The quality of the process in question is defined in terms of the process possessing specific attributes.
2. The verification of the existence of the attributes is established by specifying pertinent data to be taken.
3. A data-collection scheme must be devised.
4. Scales must be formed so that differentiation in preference can be included.
5. The dimensions of the attributes that denote quality must be weighted so that they can be combined into a set of indicators.
6. The indicators must be compared to generally accepted standards, both normative and empirical.

We want to invoke this general process to develop a comprehensive and systematic methodology for evaluating the potential benefit of alternative transportation proposals (*3*). To do this we must

1. Understand the major objectives of the groups affected by a transportation system change.
2. Determine system performance by evaluat-

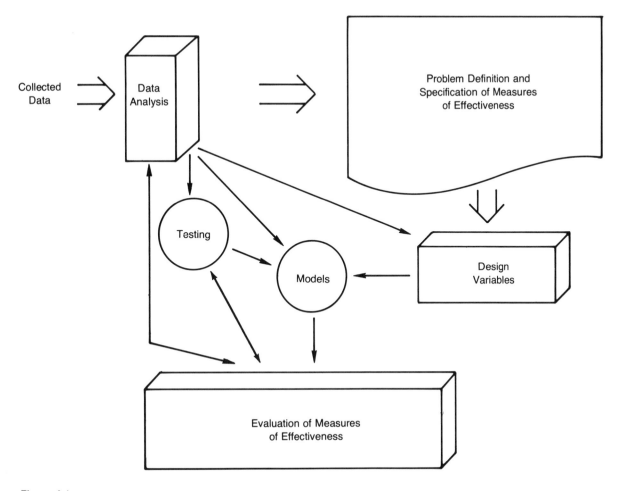

Figure 4-1

ting a number of attributes and assessing the relative importance of the attributes.

3. Analyze the outcomes of decisions directed to a wide range of mixes of transportation modes.

4. Predict all possible impacts of transportation systems changes upon urban and rural areas, and larger regions.

The evaluation methodology incorporating all these aspects must necessarily be relatively complicated; however, it is our intention to emphasize the societal effects as much as possible. The four main parts of the methodology are initial steps, trip analysis, mix of modes analysis, and network analysis. A brief description of the steps of each part is given next.

Initial Steps

a. Develop a structure of objectives for the three main interest groups: users, operators, and society.
b. Develop a hierarchy of attributes of a transportation system from the viewpoint of passenger users, freight users, operators, and society in general.
c. Choose units of measure for the attributes in a quantitative form or unambiguous qualitative forms.
d. Categories must be established to divide people into groups according to the way they perceive consequences of the transportation system.
e. For each group of step d and its pertinent attributes, a curve is prepared showing how that group perceives each attribute as a function of its physical measure.
f. In step b attributes should have been identified in a form so that they are worth independent. Any attribute interdependencies should be eliminated from step e. If the attributes are independent, the possibility of arranging tradeoffs is realistic.
g. Each interest group has specific attributes, and the relative importance of each attribute in the group must be established by assigning a weight to each attribute.

Trip Analysis

h. Using the performance measures of step c, estimate physical performance values for each trip alternative.
i. Using the curves of step e, determine the worth scores for each attribute for each group.
j. Multiply the values from step i by the weights from g to obtain weighted worth scores for each attribute.
k. Obtain an overall performance measure by summing weighted worth score from step j for attributes related to convenience, comfort, safety, and so on of passengers. Do not consider attributes related to time or cost resources.
l. The measures of performance, user time, and user cost must be combined into a single measure. One combined value will be in terms of a worth measure, and the other combination in terms of dollar cost.
m. Steps i through l must be repeated now for other passenger groups and freight commodity groups.

Mix of Modes Analysis

This section of the methodology determines the totality of uses of a given origin–destination pair.

n. Summarize the information from the trip analysis for each alternative total transportation system fro a given origin–destination pair. The result will be time, cost, and performance summaries for alternative mixes of transportation modes.

o. For each user group and for all combinations of transportation-system mixes, use demand inputs to compute incremental benefits.

p. In this step, for the number of people affected by an attribute from step b, we want to do an evaluation for a combination of modes instead of a single mode. Similar kinds of worth measurement and summation will be done for this condition.

q. The cost inputs are incorporated by identifying the total investment and operating costs, determining the pattern by which operators and agencies incur the costs and obtain the funds, and, if necessary, allocating all costs to the recipient groups.

r. Determine the worth of benefits and costs to any group as a function of the time span of interest.

Network Analysis

s. Repeat all previous steps for all origin–destination pairs in the transportation network.

t. Combining the results from each group, rank all the transportation systems in terms of their effectiveness.

u. Rank the systems again by comparing effectiveness with the system cost.

v. After careful examination of all system performance measures, make recommendations for improving overall benefit through redesign of systems characteristics and system mixes. This could be done by tradeoff and sensitivity analysis.

The long list of steps just reviewed is an example of a particular evaluation methodology based on the six general steps given at the beginning of this section. Another by-product of the methodology review has been the identification and exposure of a variety of topics, issues, and measurements, which we may now expand upon.

A critical initial factor in evaluation is the identification of attributes. With respect to an urban transportation system, the list of attributes is seemingly endless. As an example of how a hierarchy of attributes may be started and expanded, consider the following example. If we start with the premise that an objective of the urban transportation system is to satisfy user and societal interests, then we can expand from that point. The users may be identified as passenger interests and freight interests, and further subdivision of passenger interests yields travel time, cost, safety, comfort, and convenience. Continuing one step further in the attribute hierarchy, we can consider the attribute of "travel convenience" as consisting of fifteen attributes, each with the physical dimension of time (2):

1. Responsibility to arise early enough in the morning.
2. Responsibility to allow extra time for unforeseen contingencies.
3. Responsibility for searching or renting a private access vehicle.
4. Loading baggage.
5. Operating private vehicle.
6. Maintaining private vehicle.
7. Parking or returning private vehicle.
8. Unloading baggage.
9. Walking to and from other than the line-haul vehicle.
10. Waiting in queues.

11. Checking baggage.
12. Walking to and from the line-haul vehicle.
13. Sitting passively on board the line-haul vehicle.
14. Claiming baggage.
15. Retiring at night.

It should be clear from this sample of the attribute hierarchy that a long and detailed list could be developed. For our purposes a summary is sufficient and the four basic classifications and their attribute categories are given next.

Society	*Operator*
Human physiology	Profitability
Right-of-way issues	Survivability
Economic	Operational freedom
Urban form and design	
Sociological–political	
Scarce resources	

Freight Users	*Passenger Users*
Shipment cost	Travel time
Shipment time	Travel cost
Environment	Safety and security
Reliability	Comfort
Convenience and flexibility	Convenience

Next we want to examine ways to measure attributes, and for our sample area we will look at the classification designated "Society." A valid generalization is that among the most important factors in urban transportation systems are the societal issues, and unfortunately these are the most difficult to understand, quantify, and deal with. Clearly, then, societal attributes are worth our serious consideration. The hierarchy of societal attributes appears in the following lists.

Scarce Resources

Radio reception: interference, spectrum allocation
Electric power capacity
Ecology: air pollution, water pollution, biological equilibrium, climatic equilibrium

Sociological–Political

Individual life style: Leisure time and activity patterns, autonomy
Community involvement and communication
Political feasibility
Governmental form and intergovernment relationship
Tax policy: community tax base, tax incidence
Government-supplied services: police, fire, refuse disposal, education

Metropolitan Form and Design

Population distribution: urban areas, new towns, regions, megalopolitan fusion
Land-use distribution
Housing: quantity, quality, and living density
Design: urban structures and their layout

Economic

Employment
Personal income
Industrial distribution

Right-of-Way Issues

Partitioning
Displacement
Construction inconvenience
Property values

Human Physiology

Air pollution
Noise pollution
Accidents in nonusers

The societal attributes and their dimensions are as overwhelming as we might suspect. The societal impact of an urban transportation system is enormous, and as smaller order impacts are examined, it becomes clear that they interface with effects of other urban phenomena. In order to keep the measurement problem manageable, criteria must be specified that will place limits on the number of effects primarily attributed to the transportation system.

The list of societal attributes represents one model and means of determining transportation-system impacts. To be more specific about these attributes, we need to define them carefully and suggest ways they can be measured.

Pollution is an urban issue receiving some contemporary emphasis. If we examine air pollution, we must separate the transportation system pollution effects from other pollution effects, such as other transportation modes, industrial sources, and population concentration. A pollution model of the area will allow separation of the pollution sources. Following this difficult task is an even more difficult one. The air-pollution levels must be related to health effects. Unfortunately there is no general agreement on pollution-related health-effects evaluation. The most helpful material in the literature suggests measurements in terms of different classes of symptoms, such as (1) no effect, (2) a feeling of dirty skin, (3) stuffy nose, (4) eye irritation, (5) coughing, and (6) asphyxiation. These categories must be considered with respect to time of exposure and concentration of particular pollutant involved. In addition, air pollution is associated with location and can be related to socioeconomic groups.

There are several aspects to the right-of-way attribute of transportation systems. The division of communities by a right-of-way is called *partitioning*. Partitioning can be measured by the number of communities partitioned and the length of the partitions. *Displacement* refers to the forcible movement of households, firms, and institutions from the right-of-way. Typical displacement measures are number of people, acres of land, and building square footage. Measures of *construction inconvenience* include time delay of pedestrians, dirt impinging on nearby residents, and reduced industrial production during construction period. The change in *property values* is represented as a desirable or undesirable income redistribution.

The measurement of dimensions associated with sociological-political attributes are likely to spark the greatest amount of interest and controversy. The components of the *leisure time and activity pattern* of individual life style include the demand for cultural, entertainment, and recreational facilities and the set of activities centering around the household. A way to measure the effect of a new transportation system on recreational and cultural facilities is to show the change in the attendance. Surveys of household activities would provide data on the second dimension. Here again a variety of causal agents are in operation, and isolating the transportation system impact may be difficult. The demand for these leisure activities must be classified by socioeconomic group.

Community involvement and communication is an attribute that measures the level of personal contact among individuals in a community and the rate of participation in community affairs.

Because transportation systems influence population distribution and movement, which in turn affects personal contact, this appears a reasonable dimension to consider. However, at this point its measurement and use is speculative. The necessary empirical research has not yet been done.

The proposal of an alternative transportation system has a number of political implications. The attribute of *political feasibility* attempts to measure these. The necessary funds and legislation required to implement an urban transportation system causes a great deal of pressure on a number of political institutions. Reference (3) suggests that a mail polling of uninvolved political experts with the Delphi technique would provide an estimation of political feasibility. Again this dimension is speculative, and specification of measurements requires further study.

The attribute identified as *governmental form and intergovernmental relations* attempts to account for the impact of transportation authorities, and planning agencies associated with transportation systems, on existing government. A variety of events may occur which can affect government. Communities will be joined, populations increased, demand for a regional government may occur, and the role and coordination of municipal government can be influenced.

The *tax policy* attribute has more obvious and direct connotations. Some dimensions of the tax effect on local communities include the withdrawal of real estate from the tax rolls, the change in the ratio of business to residential property, the change in the socioeconomic character of the community population, and the effect of the transportation system on the relation between the population a community can tax and

the population a community must provide services for. This last point is very important, that is, the cost of services that must be provided for nonresidents and nontaxable residents, such as welfare rolls. Communities from which population and industry move because of transportation facilities elsewhere will have a tax impact. Another aspect of this attribute is a measure of the incidence, by socioeconomic group, of the taxes required to construct the transportation system.

Government-supplied services is the last attribute of the sociological–political category. In this case the contribution of transportation system change is measured for public services such as police protection, fire prevention, refuse disposal, and education. If there is a change in the level or socioeconomic distribution of population due to a transportation system change, there will be an impact on all public services which must be evaluated.

Thus far we have taken a broad overview of the transportation system evaluation methodology, and a somewhat closer look at the societal attributes in the sociological–political category. We are going to pull together these and other parts of the complex methodology by the review of a complex example that is given in reference (3). The example deals with the evaluation of a multimodal transportation system in the Washington–New York corridor. The example is contrived; however, it can be viewed as a proposal for a new transportation system. The basic issue is that the Washington–New York corridor is an area that will experience substantial population increase in the next decades; large transportation demands will occur; the coordination of the development of the region and the interdependent transportation must be considered. In order to proceed with this example we must make an assumption regarding the desired development of the corridor. We will assume that population growth will be contained in the central part of the corridor, and that industrial and population growth will be diverted to the eastern and western parts of the corridor. Based on these assumptions a development plan must be formed that will allow the population distribution evolution to occur as assumed. This plan includes a range of governmental and private actions, including an evolutionary transportation system, to allow the redistribution of several million people in the region over a 30-year period.

The part in which we are interested is the proposed transportation system which will be introduced into the region in a step-by-step fashion over a 20-year period. The elements of the total transportation system in its final form are:

1. A network of guideways: Using existing rail or freeway right-of-ways at ground level, the guideways will be capable of carrying individual vehicles, vehicles formed in trains, or ferry trains.
2. A high-speed train: Powered by conventional motors, this train will operate at speeds up to 200 miles per hour carrying passengers and including flatcars for vehicles and freight.
3. Unloading, switching, and terminal facilities: Flatcars and individual vehicles on intercity travel will use sidings at stations for loading and unloading freight and passengers.
4. Self-propelled vehicles: For intracity travel these small electric cars and trucks are capable of speeds of 50–70 miles per hour and a range of 100 miles.

The multimodal transportation system just described is intended to provide faster service between major cities in the corridor than would an evolutionary freeway system. Following are some of the comparisons between a multimode (MM) and a freeway (F) system.

1. Average speed on MM is 90–100 miles per hour; on F, 60 miles per hour.
2. The operating cost of MM will be about $0.05 per passenger mile versus $0.10 for F.
3. Construction cost of MM system would be about half that of freeways.

These gross features of comparison are too limited for real evaluation of the alternatives of multimode versus freeway transportation systems. Assuming that we have applied the several steps of the evaluation methodology reviewed earlier, we shall now summarize the results of measuring the societal attributes of the multimode (MM) versus the freeway (F) system. This "soft" version of the resulting data represents the form of the evaluation results that would be provided to a decision maker.

Societal Attributes of the Multimode and Freeway Systems

1. Human Physiology

F system: Direct air pollution high near right-of-way, moderate to high in cities, and moderate in regions with many people affected. The concentration of industry and population will indirectly cause high pollution levels in urban areas. High accident rate.

MM system: Direct pollution concentrations low in all areas, and indirect pollution moderate. Accidents moderate to low on guideways and low on other subsystems.

2. Partitioning

F system: Freeways will generally require 200 feet of right-of-way on the surface, and the partitioning will be moderate.

MM system: 60 feet of right-of-way either underground or elevated; uses existing right-of-way, therefore little or no partitioning.

3. Displacement

F system: 20 acres per lineal mile with 90 acres per interchange, 200 people displaced per lineal mile.

MM system: 8 acres per lineal mile with 90 acres per terminal, 80 people displaced per lineal mile where no previous right of way exists.

4. Property Values

F system: Value increase distributed along freeway route, $40,000 to $60,000 per acre near interchanges and $500 per acre in western area.

MM system: Value increase concentrated at terminals and in discrete cities. $80,000 to $100,000 per acre near terminals and $1000 per acre in western region.

5. Employment

F system: Large construction employment with many workers coming from western area. Seasonal unemployment largely unaffected.

MM system: Large construction employment with a moderate number coming from western area of corridor. Seasonal unemployment reduced by transportation system and by industrial development.

6. Personal Income

F system: Moderate stimulation of per capita income with the largest industrial production and per capita income concentrated in urban centers.

MM system: Moderate stimulation of per capita income due to construction but large stimulation due to total development plan. Increases in production and income spread evenly over region.

7. Industrial Distribution

F system: The farming character of the western region is retained except in a few centers.

MM system: Farming displaced in relative importance by manufacturing, retail, service, construction, and recreation industries.

8. Population Distribution in Urban Areas

F system: Population spreads along routes, and large urban centers are encouraged to expand.

MM system: Population concentrated at terminals with outward spread of city inhibited. New cities are created and small to medium-sized cities expand. Population is concentrated at terminals and spread of city is inhibited.

9. New Towns

F system: A group of satellite communities dependent on center city will grow in a distributed pattern.

MM system: Dependent and independent cities will be encouraged with spatial separation between themselves and center city.

10. Regions

F system: Small population movement from central to west region of corridor. There will be homogeneous socioeconomic population distribution in medium to low density along freeway routes and near large cities.

MM system: 5 million people will move from central to west region with the population concentrated in discrete areas along the route.

11. Land-Use Distribution

F system: Low-density usage in western region; little recreation land or open space near cities; rapid conversion of agricultural land to subdivisions; and high use of land for transportation purposes.

MM system: Most of western area will be used for medium-density residential housing; large amounts of open space in cities with open rural lands surrounding cities; moderate use of land for transportation purposes and reduced development of subdivisions.

12. Architecture and Urban Design

F system: Modest opportunity for design innovation in system with low visual image.

MM system: High opportunity for design innovation and strong visual images from high density and terminals.

13. Leisure Time and Activity Pattern

F system: Leisure time is reduced for three-quarters of the people; recreation areas move outward with large distances between them; a great deal of movement between cities but not in cities.

MM system: Half of work force live and work in same community with an open space and recreation areas close by; recreation activities are concentrated and there is a great deal of movement in cities.

14. Community Involvement and Communication

F system: Limited involvement because of population spread; only moderate communication between communities in western region.

MM system: Well-defined communities with high accessibility causes high involvement in cities and moderate communication between them.

15. Political Feasibility

F system: Results in the least change in the power interests of groups and bureaucracies.

MM system: Major changes in power structure with some uncertainty about the nature of the changes; few interest groups or bureaucracies exist to support a multimode system.

16. Tax Policy

F system: A moderate amount of land is withdrawn from the tax rolls; some towns will have a high tax base, others low; some tax disparities will develop.

MM system: A small amount of land is withdrawn from the tax rolls; there is a balanced tax base for all communities; few tax inequities will develop.

17. Public Service and Utilities

F system: Public facilities and services must be distributed over a large area; slow but easy access for auto owners; difficulty will be experienced in responding to emergency calls.

MM system: Concentrated public facilities and services will be available with easy and fast access for all groups; emergency responses will be facilitated.

In retrospect, this long and complicated evaluation methodology provides a framework for simultaneous consideration of a broad selection of significant dimensions of a complex multidimensional phenomenon. Serious difficulties exist in some of the actual measurements; however, even their imprecise inclusion is very important. It is clear that transportation systems have an incredible impact on urban affairs. Their effects are difficult to separate from basic social, economic, and political impacts. The example just reviewed, though contrived, exemplifies a contemporary transportation problem, that is, the extension of the freeway system versus development of a multimode system. In the next section, another alternative will be examined.

Demand-Responsive Transportation System

The U.S. Department of Transportation has predicted that transportation demands in this country will at least double during the 1970s (4). By any measure it does not appear feasible to meet this demand in a traditional manner. Issues of pollution, ecological damage, and land-use requirements seem overwhelming. General awareness of these difficulties has caused the development of many alternative urban transportation schemes. Dominating this thinking are a number of mass-rapid-transit systems. The common factors all systems possess are *fixed routes,* and other considerations are small or large vehicles; elevated, surface or subterranean rights of way; rail, guideway, or tube guidance; wheel or air-cushion suspension; and electric or pneumatic motive power. In spite of highly touted other advantages, mass-rapid-transit systems cannot provide the comfort and convenience the motoring public has come to expect, and 70 percent more cars are expected on our highways by 1990.

The concept of mass transit is, of course, an increase in the size and a reduction in the number of vehicles with accompanying decreases in unit costs, roadway requirements, and pollution consequences. A natural consequence is that frequency and/or geographical coverage of the service must suffer. In addition, the rights-of-way are not used by other traffic, and often auxiliary travel has to be utilized at either or both ends to complete the trip. In an attempt to deal with these principles and still lure more travelers away from their cars, a new breed of transportation systems, known as "demand-responsive" systems, has been proposed. This new system attempts to combine the door-to-door convenience of taxis with the economies of conventional buses. The intracity electric vehicles of the previous example could be operated as a demand-responsive system. The vehicles operate like taxis in that they go anywhere within a service zone in response to telephoned requests for service. As they proceed along their way they may call for and deliver a number of passengers who have somewhat compatible itineraries. The route continuously varies, and the tradeoff is to see that no customer gets so irritated either because his trip takes so long or because he has to wait too long to get picked up that he withdraws his patronage. Beyond a certain level of service, the complexities of this system would be hopeless without the memory and organizing abilities of the digital computer.

The technical feasibility of demand-responsive systems is certain; however, their economic position is unsure. Experiments conducted in Peoria and Decatur, Illinois; Flint, Michigan; and Mansfield, Ohio, have given promising results.

An important psychological factor which emerged in studies is that anonymity is an important facet of both public and private transportation. It appears that the average citizen is willing to travel alone in a car or with fifty or more strangers in a bus, but to be with one or two strangers in a demand-responsive system vehicle is threatening. This observation gives some support to the notion that affluence is used frequently to gain autonomy, which, in turn, is used to decrease intimate contacts. The familiar "alone in a crowd" phenomenon is an example. The psychological factor underscores the fact that the transportation capacity we need for many decades is available now and being wasted because the automobiles presently on the road carry an average load of much less than half their capacity. The possibility of using this capacity in a demand-responsive system has been dreamed about, but resistance via a variety of mechanisms would be substantial. Psychological and political obstacles include fear of sharing rides with strangers, the opposition of the transportation industry, and the implications for insurance and for driver liability. In addition, the idea of cooperative solutions to community problems appears to run counter to a cultural competitive spirit in which we use technology to gain relative advantages over each other.

We have attempted to show, once again, how the urban transportation system shapes and is shaped by the social structure within which it operates.

Air Transportation Systems

In this final section of the chapter on urban transportation we will look at an intercity process, air transportation. Clearly, the air transportation system must link up with the urban transportation systems we have been discussing. There is little disagreement that the demand for air transportation will increase substantially in the next 10 years. The issue we are to study, briefly, is the systematic planning for this growth by considering the problem and some of the alternative solutions. To allow us to be more specific in the presentation, we will use the New York region as a case in point (5).

The Port Authority of New York has estimated that in 1980 there will be three times as many air passengers as there were in the late 1960s. Planning for adequate capacity raises numerous technical and economic issues. Seven possible ways are available to add to air transportation capacity.

1. Rationing of peak capacity: By various means, air transportation activity would be

spread throughout the day rather than being concentrated during particular periods.

2. Larger aircraft: A trend to larger aircraft exists, and this could be encouraged by suitably arranged landing fees.

3. More landings from present runways: The historical improvement of runways must be predicted for planning purposes.

4. Development of peripheral airports: Short to medium-length flights could be handled from several smaller peripheral airports.

5. Intercity STOL and VTOL aircraft: Short and Vertical Takeoff and Landing aircraft could take over all short flights and relieve some of the pressure on major airports.

6. Expand existing airports: Of the three major airports in New York, John F. Kennedy has the potential for expansion.

7. A fourth major airport: Construction of another major airport is a frontal attack on the capacity problem, to say the least. Land acquisition and construction lead-times are substantial.

Variations or combinations of these seven possibilities could have an impact on the future air-transportation-capacity problem. In addition, the development of high-speed ground transportation would transfer short-haul passengers out of the air market.

A reasonable way to view the alternatives would be by doing a cost-benefit analysis. A rigorous study would be very difficult, however the cost-benefit concept is generally helpful in planning. If we assume that most of the benefits of air travel are experienced by the users, then the money and time invested by the travelers become significant dimensions for planners and decision makers. The costs imposed on the community at large must be acceptable.

Development of a model of an air transportation system is somewhat easier because of the natural limitations on its scope. The cast of characters, elements, and activities involved in the system are relatively easy to identify. An informal model exists in the content of the following statements (5):

1. Total trip time and total trip cost are the significant attributes of air travel, and these dimensions will be used to determine performance.

2. Passengers are assumed to use the airports to which they are closest in time and at which equivalent schedules are available.

3. A subsystem of the air transportation system must be the urban access transportation system.

4. A route and cost model must be developed for the various types of aircraft and trips.

5. Airport capacities must be computed using combinations of strategies given earlier.

Within the framework of this model the seven strategies for expanding system capacity can be examined. This approach for a specific city would allow evaluation of system alternatives in terms of levels of service, trip attributes, composition of demand, regional pattern of demand, effects on particular user groups, predicted peak load, and effect of cargo traffic.

Reference (*5*) considered many of these points with respect to the New York area. A summary of the authors' observations is given to conclude this chapter.

1. The most efficient way to get more long-haul capacity is to add specialized short-haul airports.
2. In the near future expansion will be limited to that provided by Vertical and Short Takeoff and Landing aircraft.
3. Even with the prospect of a fourth airport, short-term measures will be necessary.
4. Planned phased system growth is the only way that a desired long-run position can be attained.

References

(*1*) J. W. Dyckman, Transportation in the Cities, *Sci. Am.,* **213**(3), (1965).

(*2*) M. Wohl and B. V. Martin, *Traffic System Analysis for Engineers and Planners,* McGraw-Hill Book Company, New York, 1967.

(*3*) F. S. Pardee et al., Measurement and Evaluation of Transportation System Effectiveness, Rept. RM-5869-DOT, Rand Corporation, Santa Monica, Calif., Sept. 1969.

(*4*) D. W. Kean, Humanistic Technology: A Contradiction in Terms, *The Humanist* (Jan.-Feb. 1972).

(*5*) H. S. Campbell et al., Alternative Development Strategies for Air Transportation in the New York Region 1970–1980, Rept. RM-5815-PA, Rand Corporation, Santa Monica, Calif., Aug. 1969.

CHAPTER 5

Regional Analysis

Analysis of Cities and Regions

In this chapter we are concerned with methodologies and concepts which contribute to the analysis of cities and regions, and the process of planning cities and regions. In the most general sense, regions are subdivisions of nations. By more elusive criteria we could define a city, and if in a given region a city was the major influence, then we could say we have a city region (1). Clearly, this language is not very precise and will include urban components we have discussed already. There are four basic components which a city or region will possess. The first component is all mobile objects such as people, vehicles, and goods. Next, we list all immobile objects, such as buildings and transportation facilities. These components are referred to as the *physical infrastructure*. The land itself is a component, and, last, all activities that take place in the region form the fourth basic component. There are social and economic activities, and a wide spectrum of interactions, including traveling and shopping. This particular way of describing regions is the systems identification approach, in which entities and attributes are determined at the outset.

To continue the specification of a region, two scale factors must be set. The *spatial* dimension must be determined so that components requiring spatial specification can be located. The *sectoral* dimension sets the level of detail to which each component should be classified. For example, how finely should the total regional population be broken down?

The simple ideas presented thus far allow a variety of regional structures to be formed. Possible aggregations of entities and activities include those given in Chapter 3. The spatially distributed population could, for example, be aggregated by activity and location, such as residential location, work-place location, shopping, use of public services, and social activities. From this population activity by location, the economic activity could be classified, as commerce, construction, manufacturing, retail, public services, social and other services, and public utilities. Each of these categories would, in turn, lead to specific land uses.

The most obvious limitation of the elementary regional structure presented thus far is its inability to deal with dynamic processes such as obsolescence and urban renewal. Our simplified model, containing regional entities, attributes, and activities, is a static model, and we must find ways to modify it so that on-going processes can be accommodated. In addition, the regional structure presented has been described but not explained.

Hypotheses, concepts, and theories that contribute to an understanding of the evolution of regions have been obtained in a variety of ways. The most elementary approach carries out a statistical analysis of the static model for several points in time. The statistical regularities determined in this way lead to elementary theories. For example, the spatial analysis techniques developed by geographers have been used to formulate regional models. Among the tools used are factor analysis and regression analysis.

Our knowledge of the theories of cities and regions is highly imperfect. There is no universal body of knowledge which can be called upon and is generally accepted. The different disciplinary approaches to cities and regions have led to a particular emphasis for each case. Demography is concerned with population structure and movement at spatially aggregated levels; sociology deals with population activity; political science analyzes governmental activity; economics considers the ebb and flow of goods and services; and geography is concerned with spatial distribution.

This disciplinary mix has lead to studies being conducted by multidisciplinary teams of the type we have discussed so often. The hope is that central concepts of cities and regions will emerge somewhat independently of the focus of the separate disciplines. The integrative approach of systems analysis is a strong contributor to the discovery of general propositions associated with cities and regions.

We discussed many urban models in Chapter 3 and pointed out the importance of the comprehensive Lowry model. Modeling progress has been made under other headings.

Good models of population structure exist at spatially aggregated levels if not too great a degree of sectoral detail is required. Matrix models can be used at the national, regional, and city levels to calculate populations by age and sex, subject to assumptions about birth rates and migration (2). Elementary residential location models exist but have difficulty incorporating appropriate value structures for the inhabitants. Many economic models exist and a review of this literature is given in reference (3).

Planners require that models of land use and infrastructure be formed separately, and this has been difficult to do. Most physical infrastructure changes its use during its lifetime. The most highly developed regional submodels are the transportation models reviewed in Chapter 4. Although the quality of the transportation models is high, since they must connect with other regional models of lesser quality, the integrity of the total model is less than desired.

Regional Planning and Theory

Abstraction and generalization of theory and technique in relation to regional planning is very difficult (*4*). Applied regional economics consists primarily of models based on the concept of multipliers. These models range from the simple economic base model to complex social accounting and econometric simulations. Economic models have projected jobs and assumed that the population would adjust itself to them, and demographic models have projected population trends but left the economic question unanswered. These models are being brought together. There is evidence that jobs and people are interdependent; that is, people follow jobs, and jobs follow people. In many cases the basic questions are not related to issues in the models we've just mentioned; rather the crucial need is for regional economics to suggest ways to identify and attract economic activities to a given area.

Geographers are contributing to regional theory by attempting to identify the basic dimensionality of socioeconomic structures through the use of graph and set theory, spectral analysis, and communication and general systems theory.

The data base necessary for theory development has improved in the form of highly detailed censal information often available on computer tapes. The computer itself is substantially improved, and highly skilled researchers are being attracted to this work from the several relevant disciplines.

Evaluative techniques in regional theory are much less developed than the methods of analysis and projection. Cost-benefit analysis, social indicators, budget programming, and efficiency analysis are among the techniques being applied in the face of many practical difficulties.

Thus far we have mentioned the "hard" or quantifiable aspects of theory development. The nonquantifiable aspects of regional theory, including social, political, and institutional aspects, have lagged by comparison. This has to do with the need to study the relations of the regional theorists to their clients in the process of analyzing and planning, and, later, in how the plans are used in the political process of decision making and the actual process of implementation.

In the United States there is no commonly accepted theory of planning (*4*). However, a great deal of explicit regional planning goes on, even though decisions for regional development take place in a complex and shifting environment. Based on some of the objections to a systemic approach given in Chapter 1, there is relatively little interest in master plans, and if they exist, often they are not taken seriously. At this point in time, there is emphasis on analysis and projection using the models we have described, on voluntary coordination, and, to some extent, on evaluation methodologies.

As a final comment in this section on regional planning and theory, a way of inferring just what topics are appropriate to include lies in a careful examination of the following list of areas of study suggested for a student interested in regional planning (5):

1. Economic location theory.
2. Central place studies.
3. Analysis of metropolitan and intermetropolitan ecological structure.
4. Regional economic development.
5. Urbanization processes and the role of cities in the historical development of regions.
6. Theory of resource use.
7. The spatial structure of decisions and political authority.

Location Theory

The activity of planning covers a broad spectrum of topics, but its intent is always to *promote better performance* of the physical environment in accordance with a set of broad aims and more specific objectives set out in a plan (6). Such a plan constitutes regulation and control on growth and requires an understanding of the processes of change. If a developer is refused permission to establish a new subdivision because the subdivision would violate the public interest, this requires a clear definition of public interest and an understanding of the specific repercussions from the housing development. Any new services, one-way streets, airports, and so on that may arise as a consequence must have a predicted time and spatial location and an indication of the magnitude of the activity. The relevant principles to be invoked are contained in *location theory,* a still-evolving body of thought and knowledge.

First theories of location appeared in the early 1800s, and work continued until in the late 1940s when two common factors had emerged from all the studies that had been conducted. One concept was the idea of an equilibrium condition in which change was explained as a reaction to an outside disturbance, after which a new equilibrium would be reached. The second concept said that any locational decision, regardless of the decision makers, was made rationally in order to choose the best location for their activity.

These early results have been subject to many attacks. Change is an ever-present ongoing process for which the idea of an equilibrium condition seems inappropriate. By careful definition, brief periods of time might be viewed as in equilibrium. For example, changes are small in many sparsely settled regions of the world.

The rational-decision-making concept assumes that all relevant information is available and examined, and that an optimal decision maximizes the net benefits to the decision maker. Von Neumann and Morganstern pointed out that for a large proportion of decisions, less than the best possible outcome is accepted. They related decision making to strategies in game playing. The effect has been that decision theory has placed a growing emphasis on gaming, probability, and random processes.

Land-use activities that could be predicted from location theory were in crude form at the end of the 1950s. Highly complex urban transportation systems were being planned on these very shaky land-use patterns. We have already discussed the *interdependence* between transportation systems and locational behavior.

In the last two decades a wide variety of individuals and disciplines have contributed to the growing literature on location theory. Four major themes seem to reappear again and again:

1. Dynamic rather than static models are needed in order to obtain continuous rather than discontinuous analysis.
2. The system description must take account of activity linkages where change in one element has the effect of altering the environment for change in another element. Thus interdependence and feedback effects in locational behavior must be understood.
3. A probabilistic rather than a deterministic view of human interaction must be taken.
4. The interrelation of policy, proposal, and action, arranged to lead sequentially from one to the next, must be understood and implemented.

To give some specificity to this discussion, the next few paragraphs will give a brief review of classical location theory (7). The most classic problem in the literature is that of choosing a location for a factory. The statement of the problem assumes that material sources, location and size of markets, requisite quantities of material per unit of product, and the transport rates are known. With everything except the transport costs held constant, the best location is clearly the one that minimizes transport costs. If we consider the transport costs at a variety of locations, we can model the variation of the costs as a surface, the height of which at any point is proportional to the transport costs at that point. The lowest point will be where the smallest costs are; however, there will be pits at each material or market location.

These discontinuities at the source of material or at a product market occur because of the avoidance of terminal costs such as loading or unloading, the need for insurance, or arranging for shipments. Transportation savings are possible also at ports or any place a transshipment occurs. If material is taken from a train to be loaded on a ship and processing can be done at this point, there will be a substantial transportation savings. Location theory recognizes that in addition to transportation costs, additional savings can be experienced because of lower wage rates, preferential tax treatment, or other manufacturing advantages.

The systematic consideration of transport and production costs at all possible locations and choosing the lowest is not a true economic model. A more meaningful and more complex location model would account for factor substitution, the elasticity of demand, economies of scale, and the consequences of alternative pricing policies. Factor substitution refers to a change in a manufacturing process to account for the fact that diverse locations will result in different delivered prices for materials.

The location theory presented thus far views the location of the factory as responding to pulls from markets or materials. The best point for location will be the one where the several forces balance each other and equilibrium is achieved. Processes that lose weight, such as the reduction of ores, are likely to be pulled toward the source of the material, and processes that increase in weight or bulk, such as beer or automobiles, will be pulled toward the marketplaces.

Classical location theory is, for the most part, based on static models, and what is needed are the dynamic conditions for developing space economies. Additional factors include differences in information, which involves uncertainty, knowledge of opportunities, and technical and managerial capacity and supply. Industry that has any dependence on close contacts or on any externalities is likely to locate near existing centers.

The location theory just reviewed applies equally well to projects where the location decision is the responsibility of government. One of the complex additional factors to be considered is the appropriate rate upon which to calculate interest costs on public investment. The difficulty centers around a possible difference in the interest rate applied to capital costs and the discount rate applied to future profits, and there is no unanimity on the proper rate.

A cornerstone of location policy has been the problem of city size. Most nations believe their cities are too large, and therefore they pursue policies of decentralization, particularly for manufacturing industry. The basis of this belief is the idea that a set of costs increases rapidly after a city reaches a certain size. Among the costs are traffic congestion, water and sewage disposal, shelter, policing, and other social costs. A counterposition proposes that productivity may also increase with size. Factors contributing to this argument are the facility of communications, the availability of information and specialized services, the richness of interindustry linkages, and the principle of massed reserves. (7).

In the determination of a city versus a rural setting, other more elusive factors are worthy of examination. Sometimes a rural industrial site may be viewed as an investment in human resources; that is, the nearby society will be transformed with new attitudes, new awareness, and new patterns of behavior better suited to economic advancement. We may also expect that such a project will promote knowledge about the location and help integrate it into the information web of the more advanced sectors of economic activity. These are important aspects of location, but they are difficult to evaluate.

Policies of the regions themselves will have an impact on location policy. Capital may be made cheaper in some locations of the region, a subsidized transport rate may be provided, or certain locations may be forbidden or promoted by government decree. The points strengthen the need for a regional plan where policies just mentioned would be contained, and information about the region would be propagated in this way. In effect, regional planning provides a sounding board and outlet for local as well as regional voices.

Location Models

Models based on location theory have been made possible and useful because of the application of mathematical models such as optimization techniques. In turn, the mathematical models have stimulated advances in location analysis (8). The results of the systems analysis of the region provide the decision maker or planner with a spectrum of results from which they can select solutions that seem to meet the needs and demands of their particular region.

Location models exist for both the private and public sector. Both types of models are alike in that they share the objective of maximizing some measure of utility to the owners, while satisfying constraints on demands and other conditions.

More analytical effort has been directed to the private sector. Private-sector location analysis confronts the issue of a tradeoff between the cost of building and operating facilities to meet demands for a product and the cost of transportation. This is the classical problem of locating a factory as reviewed in the previous section. Now we want to consider any random variations in the problem. Product demands are subject to seasonal, day–night, economic, and weather variations. Private production facilities are established for use over a long period of time, and therefore they incur the vagaries of population growth, changes in competition, changes in product acceptability, and changes in the development of technology itself. Other more subtle factors fall into the category of external effects. Included are location effects on ecology, and pollution control. In both cases an actual dollar value can be put on private activities that have public effects.

Decision making associated with locating in the public sector must cope with all the factors just mentioned, and in addition must deal with the quandary that goals, objectives, and constraints are much more difficult to quantify. Public-sector location problems include both ordinary and emergency services. Location of ordinary services facilities considers post offices, schools, highways, public housing, and water-supply facilities. A list of emergency services would include fire stations, police activities, ambulances, hospitals, and civil defense.

Several approaches are used to treat the problem of locating in the public sector. The first approach uses the objective-function method just given for the private sector; however, this is extremely difficult because of poor success in quantifying the factors of the social cost of location. A second approach examines the average distance traveled or time required by those who utilize the facilities. The problem is defined as minimizing total average distance traveled subject to a constraint on the number of facilities established. Another viewpoint looks at the demand created and attempts to maximize the efficiency in meeting the demands of the region. In all cases the optimization is accomplished subject to constraints on investment. Sensitivity analysis is run after solutions are obtained using the several objectives and constraints. If parameter variations do not markedly affect the solution, a tradeoff analysis is done. Typical units of tradeoff are

1. The decrease in average distance per additional $1000 of investment.
2. The increase in demand per additional $1000 of investment.
3. The decrease in the maximum distance per additional $1000 dollars of investment.

The practical matter of making choices of the particular objectives and constraints is most complex. For this reason, solutions may be obtained for a variety of assumptions. An even more subtle problem is the question of how to locate public facilities to influence future growth patterns.

Modeling problems in location analysis can be classified into two major structural categories: location on a plane or location on a network. An example of the characterization of location on a plane is a distance measurement according to a particular metric. The formulation for the Euclidean metric is

$$d^2_{ij} = (x_i - x_j)^2 + (y_i - y_j)^2,$$

where d_{ij} = the distance between points i and j and x_i, x_j = the coordinates in a rectangular system at the ith point. Location on a network may be characterized by measuring the shortest path from any junction in the network to any other. Many shortest-path algorithms exist.

The classical factory-location problem-solution consideration began early in this century by locating a factory on a plane between two resources and a single market. The iterative process for solving this problem developed as finding the single point that minimizes the sum of the weighted Euclidean distances to the factory location.

Define

ω_i = weight assigned to the ith point (resource, population, goods demanded, and so on)

x_i, y_i = location of the ith point relative to a fixed Cartesian coordinate system,

x_p, y_p = unknown coordinates of the factory location,

n = number of points in the calculation,

d_{ip} = Euclidean distance from any point i to the factory location.

We want to minimize an objective function z:

$$z = \sum_{i=1}^{n} \omega_i \left[(x_i - x_p)^2 + (y_i - y_p)^2 \right]^{1/2}$$

where $d_{ip} = \left[(x_i - x_p)^2 + (y_i - y_p)^2 \right]^{1/2}$. The minimized value of z that will be optimal will be determined by an iterative procedure (repeated analysis) accomplished on the computer. The requirement is to continue calculation of the values of x_p and y_p until the differences between successive calculations are negligible. The Euclidean model is suitable for both public- and private-sector problems.

Both public- and private-sector location problems can be formulated as locating control points on a network. The factory-location problem just discussed can be recast in this form. Given that we have a number of demand areas for a certain product and a number of alternative factory locations where facilities may be built to satisfy these demands. The building sites will be determined by minimizing the transportation cost and the amortized facility costs. The tradeoff is that the larger the number of well-placed facilities, the lower the distribution costs. However, as shipment costs decrease, the investment in facilities must rise. Define

x_{ij} = amount shipped from plant i to demand area j,

y_i = total amount shipped from plant i,

$d_{ij}(x_{ij})$ = cost of shipping the quantity x_{ij} from point i to j, dollars,

$F_i(y_i)$ = cost of establishing and operating a facility at site i, dollars,

D_j = demand at area j,

n = number of demand areas,

m = number of proposed facility sites.

We want to minimize

$$z = \sum_{j=1}^{n} \sum_{i=1}^{m} d_{ij}(x_{ij}) + \sum_{i=1}^{m} F_i(y_i)$$

with the following constraints:

$$\sum_{j=1}^{n} x_{ij} = y_i, \qquad i = 1, 2, \ldots, m,$$

$$\sum_{i=1}^{m} x_{ij} = D_j, \qquad j = 1, 2, \ldots, n,$$

$$x_{ij} \geq 0, \qquad i = 1, 2, \ldots, m; j = 1, 2, \ldots, n,$$

$$y_i \geq 0, \qquad i = 1, 2, \ldots, m.$$

The function $F_i(y_i)$ is frequently nonlinear and not amenable to linear programming; therefore, more powerful iterative procedures must be employed.

Regional Simulation

In previous sections we have looked at some of the principles and data used by planners in performing their major regional functions—understanding interacting located activities. In order to consider seriously the possibility of a regional simulation, several other issues must be studied.

Among those we must consider are

1. What time scale will be used for planning.
2. At what intervals regional projections will be made.
3. How predictable human behavior is.
4. How to identify the effects of different planning policies.

Two types of projections may be specified. Based on limited data, a *simple projection* may be made, but with more detailed information, an *analytical projection* can be made. In the second case the projected value is often the dependent variable derived from projections of independent variables. The relationships imply an adequate amount of data to form cause-and-effect relationships, which can be of three kinds (6).

1. Deterministic causality, in which A causes B.
2. Probabilistic causality, in which A has the probability p of causing B.
3. Correlation in which A occurs in association with B but there is no observable cause-and-effect relationship between them.

Particularly in the area of human behavior, the best we can hope for is a high level of probability. As a result, projections will frequently take the form of ranges lying between the limits of stated levels of probability.

We may view the region as a system that includes activities in spaces linked by communications in channels. To project a regional system, then, requires projection of these elements. The regional activities can be divided into the categories of productive, general welfare, and residential. These three categories establish the main thread of regional life as a dependence of the population on economic activity, and similarly a dependence of the general welfare activities on the size and nature of the population. The stronger interdependence is between economic activity and the population, and the derivation of general welfare activities comes from this. Next we will review several methods of doing population projections.

Population estimates are a primary contributer to decisions about major land uses and services. The demand for water, power and waste-disposal facilities, housing, open spaces and schools, the supply of labor, spending power available for the retail trade, the numbers of private cars to be expected, and possible recreational demands can be estimated from the projected population. The following six methods of population projection are arranged approximately in ascending order of accuracy and sophistication.

1. Mathematical and graphical methods: This basic approach uses only past population data and takes no account of change processes. The data are examined to determine if the population trend is such that the increase is by a constant amount or by a constant proportion of the preceding figure. The first case defines an arithmetic progression and the second a geometric progression. In both cases the plotting of data on appropriate graph paper yields a straight line that can be extrapolated. For other kinds of data a "best-fit" straight-line equation can be formed by the method of least squares. Clearly, these three approaches are simple and relatively crude and could not be trusted to make population projections beyond the 5- to 10-year range.

2. The employment method: If we know the past values of various activity rates we can use the mathematical methods just described to obtain future values of the rates. For example, if we know the ratio

$$\frac{\text{economically active population}}{\text{persons in working age}}$$

and the ratio

$$\frac{\text{persons in working age groups}}{\text{total population}},$$

then the product of these two ratios will give the total employment as a function of population, and a simple short-range projection is possible. Regression techniques may be used to produce the necessary activity ratios.

3. Ratio and apportionment methods: The conceptual foundation of these methods as-

sumes that changes in any geographical area are a function of those in successively wider areas. In order for a population projection to be made, a time series of the populations for the area to be analyzed and a set of forecasts for the largest area are necessary. The *ratio* method plots the population of the second largest area, the region, against that of the largest, the nation. Using methods already discussed, a curve is fitted to these points, and the curve is extrapolated to intersect the projected value for the parent area at a given forecast date. This process is repeated until projections are obtained for the study area. The apportionment method is conceptually similar to the ratio method except that more accuracy is achieved by pro-rating subarea contributions to the study area projection. A related approach expresses subarea populations as percentages of the parent area and expresses these relationships as time series. Values of future population can then be determined using regression or curve-fitting techniques.

The methods discussed thus far do not directly examine the nature of population change; therefore, the projections are weaker for longer periods and smaller areas.

4. The migration and natural increase method: This method examines past data on net migration rates, and using economic conditions makes assumptions about the patterns of future migration. Using this information, a set of patterns of future natural change would be developed either by subjective projection of past maximum and minimum rates, or by extensions from national or regional projections. The essence of the technique is to begin with the starting-date population, add the migratory data, and then add the

natural changes (birth and death rates) to complete one cycle of the selected period. Cycles are repeated to the end of the projection period. Although generally a more accurate projection method, its usefulness is limited by the fact that total populations are used with no accounting for age and sex structure. An additional limitation is incurred because the effect of the age and sex composition and the different biological characteristics of the migrant elements are not accounted for.

5. The cohort–survival method: Most government agencies use this method of population projection. This analytical method allows births, deaths, and migration to be handled separately; male and female elements can be handled separately; and results can be obtained for any age group. The steps of the method are as follows:

1. Males and females by single-year age groups are tabulated separately from census data.
2. An appropriate change for each age group of males and females is made to account for migratory changes.
3. Appropriate age-specific birth rates are applied successively to each group of women in in the child-bearing range.
4. Resultant births are divided into males and females and adjusted for mortality in the first year.
5. Age-specific survival rates are applied to each age group of males and females to estimate the numbers who will survive to the next year.

These five steps are repeated until the projection date is reached. At any point in this method special adjustments to natural and migratory changes can be made. Here, as always, data on these changes, especially migratory, may not be easily accessible.

6. Matrix methods: The use of matrix algebra for population projections is one of the most recent developments. The approach of the cohort–survival method is followed in a matrix format. Initial age and sex distribution is represented as a column vector, while the incidence of births and deaths is dealt with by using a "survivorship matrix." Migration is introduced by means of a set of transition matrices in which the elements represent the probability that an individual of that age group and in a specific region at any time will move to another region in the next time period. The arithmetic of the population projection is accomplished by multiplying the initial population matrix by the survivorship matrix and then adding the net migration matrix. The matrix format is very handy and appealing; however, survival and birth rates are assumed constant over time and over each age group, and errors will be introduced by this assumption. The matrix format is ideally suited to computer operations, and the potential of computing separately the four elements of change—births, deaths, and inward and outward migration—is clearly there.

All projections are important, but very difficult, and economic projections are no exceptions. Excellent projections require pertinent theory and reliable data, and in spite of a long history of effort, economic projections suffer frequently from inadequacies in these two requirements. Only in recent years has there been a growing body of knowledge that could be subjected to tests of empirical verification. As before with population projection, we will now review several methods of economic projection, in ascending order of sophistication.

1. Simple extrapolation: Data on a variety of economic activities are available as a function of the independent variable, time. We would expect to find data on employment, volume or value of production, value added by manufacture, and so on, available from a number of sources. These data can be projected using the techniques of curve fitting, such as graphical, correlation, and least squares. Although this projection is easily done, it reveals no causal pathways or change-agent activities.

2. Forecasts: The variables of production and employment are linked by the variable, productivity, in the units of output per worker. An estimate of employment may be obtained by dividing output by productivity. Employment projections obtained in this way give more trend information on labor and production than the usual employment data.

3. Projection by sectors of the economy: A projection that gives more detail and fine structure is more useful than one that does not. The simplest form of projection by sectors of the economy extrapolates the past trend in each sector. This may be verified by summing the sector projections and comparing to a total employment forecast obtained by some other means.

4. Economic base methods: The economic base method applies the theory of international trade to the region. That is, it postulates that growth in an area's economy comes from expansion of all those basic activities which produce for export beyond the boundaries of the local area, and this increases its wealth and ability to pay for imports. Other economic activities exist to satisfy local needs. It is fairly difficult to identify the basic sectors of the economy and to define the local area. Surveys are often used to apportion the whole range of economic activity into basic and nonbasic categories. Projections can be made by projecting the basic activities of each sector using ratios of local to national trends; then, by extrapolating the past trend in the basic to nonbasic economic activity ratio, the forecasts of basic employment are expanded to a total employment estimate. The literature contains the following objections to the economic base method.

a. The dependence on employment as a measure ignores the effects of changes in productivity.
b. The ratio of basic to nonbasic economic activities is an unreliable measure and can be very unstable over time.
c. The absence of the right kind of data is a serious problem.

5. Ratio and apportionment methods: As in the case of population projections, local economic activity levels bear proportional relationships to levels in successively larger areas, and these ratios can be extrapolated. The ratio method considers only one area in each stepdown; the apportionment method examines all areas of the next larger unit and adjusts all contributions before proceeding to the next step. As a consequence, employment estimates are more accurate, and there is sufficient detail in the forecast that the results of local projections made by local employers, managers, and trade associations can be compared.

6. Regional input–output analysis (9): Input–output analysis is a promising sophisticated projection tool for regional planning. An input–output table which shows the details of a region's economy is the basis of input–output analysis. The table is a system of accounts which focuses on the relationships between different aspects of regional economy and how they combine to produce the region product. The layout of the table given in Figure 5-1 is simple and shows for each sector of the region what it must buy from other sectors in order to produce its own product, and it shows the destination of the finished product. In the example of the table format, three ficticious sectors are identified as X, Y, and Z.

The purchases of sector X will appear in column 1. Purchasing "domestic income" refers to obtaining the services associated with production, land, labor, and capital. A "final buyer" is one who buys a product as an entity which is not intended for incorporation in some other product.

In expanded version the input–output table of Figure 5-1 is capable of showing the details of intersectoral or interindustrial relationships of an economy. The first step in the input–output analysis is to assume that the input–output table is generally valid and not dependent upon the circumstances extant when the information was taken. With this assumption we can formalize the data in Figure 5-1 to show intersectoral dependence without respect to particular output levels. In Figure 5-2, purchases by each sector from other sectors are shown as a percentage of

Sales by	Purchases by			Final Buyers	Total Output
	X	Y	Z		
X					
Y					
Z					
Domestic Income					
Imports					
Total Input					

Figure 5-1

Sales by	Purchases by			Final Buyers
	X	Y	Z	
X				
Y				
Z				
Domestic Income				
Imports	100	100	100	

Figure 5-2

total input in each sector, so that the sum of the column entries must be 100.

Although the process in going from Figure 5-1 to Figure 5-2 is simple, the implications are not. Figure 5-2 describes in a formal way the impact of a sector on the rest of the economy, provided that the production in the sector continues in such a way that the proportions of product and domestic income used do not vary over time and with the scale of operation. The stability of trends requires that there be no technical improvements in the method of production, that producers do not respond to changes in prices of their components, and that proportions of inputs are required not to vary with scale of operations or with production context.

Figure 5-2, subject to the stated restrictions, allows the determination, for example, of the impact on the rest of the economy if sector X were to increase its sales to final buyers. The increase in the production of sector X would have an immediate primary effect of the suppliers of sector X, secondary effects as these suppliers increase output to meet the new demands, and higher-order effects as each increase in outputs causes further demands for inputs. The structure of input–output analysis is such that with the digital computer these various effects can be calculated. The final version of the tables we have been discussing lists the computation effects for these several effects. The series of three tables presents, in turn, statistics and sectoral interrelations, input coefficients, and the complex pattern of interrelationships. Theoretically it is possible to read from the third table an assessment of economic effects when one or more producing sectors change the level of output.

Clearly, these kinds of results would be very useful to regional planners. Questions of the following kind can be answered:

1. How will the economy be affected if one industry closes and another expands?
2. What would be the effect if an industry experienced substantial reductions in demand for its product?
3. Where are the greatest increases in demand in other industries likely to occur?
4. What are the consequences of a steady growth?

The usefulness of input–output analysis hinges first on the availability of pertinent statistical information, and this information varies in quality from region to region. In regions where there are marked economic differences and suitable statistics are available, the analysis is most likely to be useful. The other constraining assumption on this analytical approach requires that existing trends continue. Therefore, anything we learn about the present or future state of affairs has this implicit requirement. If input–output analysis is being used in a context that honors these assumptions, it can be the foundation of regional policy recommendations, assuming that it is used in conjunction with information such as the capacity limits of particular industries, labor supply and location, anticipated changes in demand, and so on. The following example demonstrates some of the previous points (6).

Table 5-1 represents the actual transactions in thousands of dollars between three sectors of the economy of the study area. The fourth category, designated "households," includes payment of wages, salaries, rents, dividends, receipt of taxes, purchases, private investment, and so on. Now expressing each row entry as a rounded percentage of its column total we have the data in Table 5-2. The table encodes such information as:

Table 5-1

	Sector 1	Sector 2	Sector 3	Households
Sector 1	50	30	20	40
Sector 2	70	40	100	30
Sector 3	95	50	70	20
Households	15	60	10	10

Table 5-2

	Sector 1	Sector 2	Sector 3	Households
Sector 1	22	17	10	40
Sector 2	30	22	50	30
Sector 3	41	28	35	20
Households	7	33	5	10

1. $100,000 dollars worth of output in sector 3 needs $10,000 dollars worth of input from sector 1, $50,000 dollars from sector 2, $35,000 dollars from sector 3, and $5,000 dollars from households, mostly in the form of labor.
2. Household income is supplied primarily from sector 2 in the form of labor.

Let us use the input–output method to account for the effect of a substantial increase, $1 million, in the value of the output of sector 1. The table indicates that to supply additional output of $1 million from sector 1 requires

1. $220,000 of input from sector 1.
2. $300,000 from sector 2.
3. $410,000 from sector 3.
4. $70,000 from the household sector.

However, each extra input just listed is extra *output* as far as the individual sector is concerned and must itself be produced by increased inputs from all sectors, and so on. As these adjustment cycles go on, the total extra output required will decrease, and the results will converge, usually between six and twelve rounds. These results are valid for the short term only because of the assumptions associated with Table 5-1. The input–output method, because of its conceptual strength, possesses the greatest practical potential for regional economic analysis and projection.

Subregional Model (*10*)

Thus far in the chapter we have emphasized the location and projection aspects of regional analysis. A number of ancillary topics are worth considering, and we will do this in the context of two simulation examples in this and the last sections of the chapter. The first example of regional simulation concerns the Leicester–Leicestershire study in England, in which a plan was to be proposed for the major land uses of the subregion with emphasis on the main components of population, housing, industry, shopping centers, and public transportation. The subregion included both city and county planning authorities and the forecast period was hoped to be as long as 20 years.

Initial investigation revealed that disposition, not rate, of growth was the central concern. The existing coal mining industry was contracting, and because of improved highway accessibility new fast-growing industries were anticipated. Where these industries were to locate, the availability of lands and services, and the desire to conserve areas of fine landscape were the paramount issues.

The simulation method to be utilized evolved from discussions of the study staff and depended upon studies of the patterns of change of employment, population, and land development in the recent past; theories of regional growth; and a review of various models that seemed suitable for the Leicester subregion.

A wide variety of exogenous information was supplied to the simulation process, including

1. A population forecast performed by the cohort–survival method. This forecast gave figures for males and females in 5-year age groups to 1996. At each forecast year the population was divided into white- and blue-collar workers.
2. An employment forecast performed by the apportionment method. From the population projection, forecasts of the future size of the male and female work force were obtained and these were compared to the demands resulting from the employment activities forecast determined from the Ministry of Labor data.
3. Other spatially aggregated forecasts were derived from 1 and 2 above, such as the number of children of primary and secondary school age, numbers of households, total income, durable-goods expenditure, and car ownership.

4. Any extant short-run commitments, including current development plans, current resolutions about housing demolition and replacement, shopping developments, conservation policies, and road-building schemes.

For the purpose of the simulation, the study area was divided into cells 2 kilometers on a side. Each complete cycle of the simulation moved the state of each cell forward by 5 years, with the state at the end of each cycle identified by the following state variables:

P_i = white-collar population in cell i,
P_{bi} = blue-collar population in cell i,
E_{di} = declining employment in cell i,
E_{si} = site-oriented basic employment in cell i,
E_{ci} = central-business-district employment in cell i,
E_{li} = labor-oriented basic employment in cell i,
E_{pi} = population-oriented service employment in cell i,
D_i = number of dwellings in cell i,
D_{ci} = penalty score representing the average condition of dwellings in cell i,
S_i = number of secondary school places available in cell i,
F_i = sale-area floor space in durable-goods shops in cell i,
L_i = land available for development in cell i.

The number of cells in the study area was about 600.

By simulating 5-year cycles, there was the opportunity between the cycles of adjusting both the structure and parameters of the model based on the outcome of previous cycles.

The actual simulation model was based on

development work which showed that population change in any period of 5 years is a function of employment and population change in the preceding 5-year period. Such a *lagged relationship* confirms common sense about the delay in the response of the labor and housing markets to changes in employment opportunities. The form of a relationship, based on the preceding result, and derived from the Leicester subregion was

$$\Delta P_{k\,(tn/tn+5)} = b_0 + b_1\,\Delta E_{k\,(tn-5/tn)} + b_2\,\Delta P_{k(tn-5/tn)}, \qquad (5\text{-}1)$$

where ΔP_k = share of total subregional population change experienced in zone k,

   ΔE_k = share of total subregional employment change experienced in zone k,

   b_n = coefficient of proportionality,

   $tn/tn+5$ = subscripts denoting the 5-year periods used in simulation.

In addition,

$$P_k = P_{wk} + P_{bk} \qquad (5\text{-}2)$$

and

$$E_k = E_{dk} + E_{sk} + E_{ck} + E_{lk} + E_{pk} \qquad (5\text{-}3)$$

from the list of state variables given earlier. This equation represents the symbolic model of the theory cast in the form of the identified and defined variables.

Using existing data from the periods 1951–1956, 1956–1961, and 1961–1966, the model was calibrated for the three-cycle sequence and the data fit was very good.

The next several steps describe the process of one simulation cycle:

1. At the outset, the basic employment allocations were specified according to the policy being simulated for the kth zone and ith cell. The effect of this action was to change the values of the employment state variables E_{di}, E_{si}, and E_{ci} in the subregion. Using an assumed value of workers per acre, the land availability state variable L_i could be updated.

2. By using the basic model equation and the data for the previous 5-year period, the population change in the k zone was calculated. The redistribution of the population change down to the i cells was done by considering the array representing the stock of dwellings, D_i. The following submodel was used for estimating the average physical condition of dwellings in a cell:

$$D_{ci} = b_3 + b_4 A_i + b_5 \left(\frac{P_{bi}}{P_{wi}}\right), \qquad (5\text{-}4)$$

where D_{ci}, P_{bi}, and P_{wi} were defined earlier and A_i is the average age of the dwellings in cell i and b_3, b_4, and b_5 are constants. The last term in the submodel, the ratio of blue- to white-collar workers, is a proxy term for average income. The computation of dwelling-condition penalty scores for each cell from Eq. (5-4) permitted decisions on the housing stock to be made. The effect of clearance, redevelopment, and area improvement could be determined. With this information the dwelling stock D_i, dwelling condition D_{ci}, and land availability L_i could be updated.

3. With the changes in population and housing known, the population was distributed to the cells by accounting for factors such as land-development constraints; local planning policies about

schools, services, and accessibility of employment; and so on. Another submodel defined the job accessibility surface W_i as

$$W_i = \sum_j \frac{E_j^\beta}{d_{ij}^\alpha}, \qquad (5\text{-}5)$$

where E_i is the total employment in cell j.

d_{ij} is the distance between cell i and j, and α and β are constants. Equation (5-5) requires a computer solution, but the results provide a useful guide in the manual process of distributing populations at the cell level. After the several ways of distributing cellular population, a simple check was made by confirming that the algebraic sum of all the 2-kilometer cell changes totaled the zone-wide change.

4. The new population distribution affects residential density, and again land availability L_i can be adjusted. The distribution of secondary school places, S_{si}, was also adjusted.

5. Now that basic employment and population changes have been distributed for the 5-year period, changes must be distributed in the labor-oriented and service groups, E_{li} and E_{pi}. Assuming that labor-oriented employment is more responsive to blue-collar population and using the socioeconomic ratios forecast for the subregion, the proportion of E_{li} in each cell was estimated. E_{pi} was distributed simply by prorating to the change in population.

6. The final task in the simulation cycle was to adjust the network of primary roads and public transport services to account for the new values of total employment and population.

In retrospect, each simulation cycle distributed

population, employment, and all other regional variables according to a network of theory, assumptions, data, speculations, and arbitrated expert opinion. The overall simulation was, in fact, composed of a federation of subsimulations. The group of subroutines was tailored to fit the growth dynamics of the Leicester–Leicestershire area. The possibility of developing a generalized regional model which would exist in totality on the computer seems remote at this point in time. This simulation is a learning simulation and is capable of continuous refinement and improvement. Each subsimulation is capable of change or replacement.

Regional Analysis of the Susquehanna River Basin (*11*)

The last regional model to be discussed will briefly treat those aspects of regional analysis that we have not yet encountered. Staff members of the Battelle Memorial Institute—Columbus Laboratories were asked to do a study of the economic growth of the Susquehanna River basin. The studies were aimed, in part, at developing and refining methodologies for the better understanding of regional and water resources phenomena in that basin. Much of the study has the characteristics of other regional models, with the exception of the emphasis on water resources. Each of the subregional models of the basin has three major sectors: demographic, employment, and water. The demographic and employment sectors are linked by a feedback loop that includes population, labor, unemployment, and migration variables. The portion of the model which deals with water is representative of a technical factor that might be of importance to a given region. Other examples of a technical factor are forest resources, minerals, or transportation.

The demographic portion of the model is similar to others we have discussed earlier in that factors directly affecting population change are identified as births, deaths, and migration. The population is desegregated into six age classes, which are chosen in such a way that they will have relatively homogeneous birth, death, and migration characteristics.

The rate of migration in the Susquehanna model is related to the difference between the subregional unemployment rate and an assumed long-term national rate. Migration out of the subregion increases when the unemployment exceeds the national rate and decreases for the countercondition. The variation of response by age group to unemployment will change the age class structure of the population in a region. The altered age class structure will, in turn, change birth and death rates. This feedback effect of migration makes it one of the most dynamic dimensions of population change in a region. Using migration to tie the employment and population sectors of the Susquehanna model together is one of the interesting distinguishing features of the model.

The tie between the population and employment sectors causes a substantial emphasis on employment in the model. Employment is used to specify economic activity as compared to using other variables such as income. The employment portion of the model utilizes the export-base theory. Export base refers to funds necessary to

supply the needs of a region in addition to the needs met by the funds acquired through export sales. A declining export base normally implies decreasing regional employment, and vice versa. Employment associated with firms supplying the goods and services of general industry grows at the same rate as the industry. Employment related to businesses that directly supply the needs of the consumer is proportional to the population of the region. The main dynamic feature of the employment sector is the market area demand operating through export industry employment.

The role of water resources in the development of the Susquehanna basin is of special interest to us. The study wanted to determine the river's effect on the economy, both existing and potential for growth, and, in addition, to determine the impact of alternative systems of river-facilities construction upon the economy. The water sector of the model related economic forces with water resource demands and supplies. As we mentioned earlier, this "technical sector" may take many forms and we are interested in the river system to illustrate the nature of a technical sector within the context of regional development.

The water part of the model was diverted into quality and quantity aspects to consider how either might become a constraint on the growth of the Susquehanna River basin. It was necessary to project needs to determine if the quantity of water along the river was adequate; otherwise, rising water costs would reduce the growth of the economy. Water-quality evaluation was based on identification of points of pollution in the river system. Projections were to be used to determine how levels of pollutants would change with population changes, economic activity, river flow, and levels of water treatment.

The river was simulated based on a water accounting system as shown in Figure 5-3. In the water accounting system, great care was exercised to differentiate between water withdrawn for consumption and water withdrawn for later return. The water model was in the form of equations that gave estimates of water use and flow in the basin.

The water-sector model has the ability to test the potential effects of constructing dams and reservoirs. The effect of dam construction activities on the economy of the subregion may be simulated, water-flow augmentation from reservoir release may be determined, and the effect

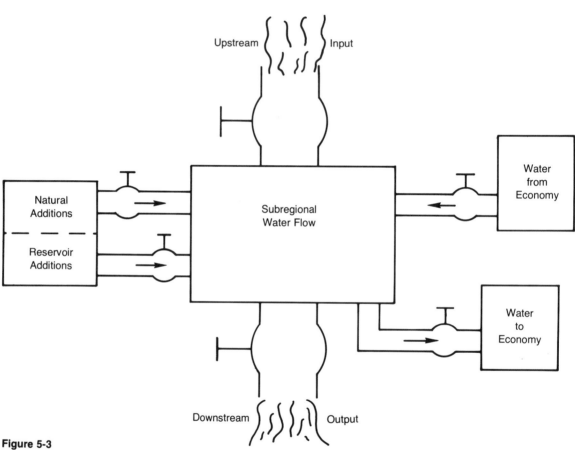

Figure 5-3

on the economy of recreation activity may be simulated.

The technical sector we have discussed is an attempt to evaluate the effects of resource management on the economic development of a region.

References

(1) A. G. Wilson, Research for Regional Planning, *Regional Studies,* **3**, 3–14 (1969).

(2) A. Rogers, *Matrix Analysis of Inter-regional Population Growth and Distribution,* University of California Press, Berkeley, Calif., 1968.

(3) H. M. Cordey, Retail Location Models, Working Paper 16, Centre for Environmental Studies, London.

(4) W. Alonso, Aspects of Regional Planning and Theory in the United States, Working Paper 87, Institute of Urban and Regional Development, University of California, Berkeley, Calif.

(5) J. Friedman, Regional Planning as a Field of Study, *Am. Inst. Planners J.* (Aug. 1963).

(6) J. B. McLoughlin, *Urban and Regional Planning —A Systems Approach,* Praeger Publishers, Inc., New York, 1969.

(7) W. Alonso, Industrial Location and Regional Policy in Economic Development, Working Paper 74, Institute of Urban and Regional Development, University of California, Berkeley, Calif., 1968.

(8) C. Revelle, D. Marks, and J. C. Liebman, An Analysis of Private and Public Sector Location Models, *Management Sci.,* **16**(11) (1970).

(9) E. M. Thorne, Regional Input–Output Analysis, in *Regional and Urban Studies,* S. C. Orr and J. B. Cullingworth, eds., Sage Publications, Beverly Hills, Calif.

(10) J. B. McLoughlin, Simulation for Beginners: The Planting of a Subregional Model System, *Regional Studies,* **3**, 313–323 (1969).

(11) H. R. Hamilton et al., *Systems Simulation for Regional Analysis: An Application to River Basin Planning,* The MIT Press, Cambridge, Mass., 1969.

**Part III
Modeling and Analysis
of Urban Subsystems**

CHAPTER 6

Community Health

The National Health Problem

The primary objective of this chapter is to develop a set of models relating to designs and decisions associated with community health systems. To place community health issues in better perspective, some discussion of the national health problem from the systems-approach viewpoint is necessary. It is generally accepted that the United States faces a crisis in its health-care system. Most of the substantiation of this statement has appeared in countless forms, including continuing coverage by the mass media. In general, the capacity of the health-care delivery system cannot cope with the demand for its services. Dramatic statistics that confirm this include:

1. The cost of health care is rising twice as fast as the cost of living. For example, the cost of hospital care has been rising 16 percent per year.
2. The average per capita expenditure for health care in this country is $300. A long-term illness can financially ruin many families.
3. The number of physicians seeing patients has declined from 103 per 100,000 Americans in 1950 to 94 in 1969. This statistic is taken to be symptomatic of the increasing difficulty of getting medical attention.

Most analysts agree that changes must occur which will allow the national health-care delivery system to attain two objectives. The first objective would be to permit every individual to receive adequate health care by making access to health care more equitable. The second objective would be to make the health-care delivery system more efficient, that is, provide greater care per dollar spent without compromising the quality of that care.

The national health-care system we have been discussing involves over 3 million employees, 7000 hospitals, hundreds of private insurance companies, and several dozen federal agencies. In one sense, however, these participants appear to form a "nonsystem," in that there is great overlapping, duplication, gaps, and wasted effort rather than an integrated system in which needs and efforts are closely related.

A number of factors inhibit the development of new methods of health-care delivery. Included are legal constraints; goals, values, and training of physicians; consumer ignorance and preferences; and lack of knowledge about the way in which health-care systems operate. And in the face of these difficulties, hundreds of new financing schemes, categories of health manpower, and ways of organizing health care delivery have been proposed and some are being tested.

The discussion thus far was intended to set the scene for stating that systems analysis has a great deal to offer to solving the national health problems. But first, before determining what the systems approach can contribute, there are two additional factors to consider. As in so many other social problem areas, the data for the health systems analyst are typically unavailable, or of very poor quality. And second, health-care delivery system problems are a mixture of scientific and sociopolitical elements. Once again, then, the not easily quantifiable social and political issues must be included in any systems study that is to have merit.

Like most social systems the health-care deliv-ery system has developed through a series of trial-and-error changes rather than by deliberate design. In a rigorous systems approach to design we would determine, in advance, the components and their interrelationships to establish a system designed to fulfill certain preassigned goals and objectives. Once any social system is established over a period of time, the identification of causal pathways, feedback loops, and transfer functions to describe processes is very difficult. Horvath has suggested analyzing and understanding complex health-care systems through a crude modeling technique (1). He proposed a primitive simulation by trying to identify the major system interactions and then formulating a set of rules for a game. Gaming was discussed in Chapter 2. Generally, practical insight can be gained from even the most primitive model and will lead to more refined models. To form the rules of a health system game, the participation of public health officials, administrators, researchers, practitioners, and representatives of the consuming public would be required. Their efforts would be focused upon obtaining a code of behavior for the different components of the system under a wide variety of stimuli. Input information would include data from the actual system and from laboratory experiments, and a consensus opinion of pertinent experts from the field. Horvath believes that the gaming approach can contribute to the following problems.

1. The costs of medical care: The cycle of the increasing complexity of medical services, followed by higher costs, which in turn are followed by increased use and cost of hospitalization insurance, should be studied.

2. The training of medical personnel: Medical schools should be studied to examine aspects such as the selection of candidates, the costs of a medical education and how it affects the quality and quantity of the future supply of physicians, the role of medical schools in setting trends, the effects of physician specialization, and the promise of paramedical personnel.

3. Hospitalization versus home treatment: Specialization and hospitalization insurance have caused increased use of hospitals. Rising costs may reverse this process.

4. Public health and sanitation: Issues in this category are environmental pollution of water, air, land, and foods; and elimination of major occupational hazards.

Since Horvath described the gaming approach to health-care-delivery-system simulation, many others have described and proposed a variety of other types of simulations. Flagle gives an example in reference (2) of the simulation study of the change of organization of services in a busy out-patient clinic. The difficulties in the clinic centered around congestion and patient delay. Data revealed that patients were spending four times as long waiting for service as receiving service. Queuing theory indicated that in principle combining some tasks to provide a parallel rather than a sequential flow would reduce this problem. Only when admissions are uniformly spaced and tasks standardized to a specific duration is a purely sequential operation suitable. The opposite condition of randomly arriving patients, including emergencies, is better served by parallel channels of flow, where each channel provides all the services of a particular classification of treatment. Figure 6-1 gives a block-diagram representation of the original situation in the clinic.

An alternative arrangement proposed by Flagle is given in Figure 6-2. Parallel channels of service require that personnel be trained and equipped for multiple tasks. The theory supports the fact that the organization of Figure 6-2 is superior to that of Figure 6-1; however, the question is how to confirm, with data, that this is true. Assuming

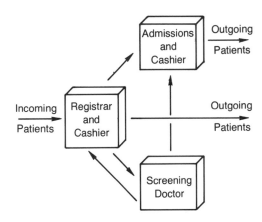

Figure 6-2

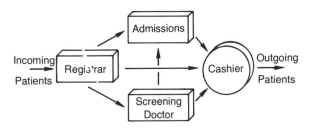

Figure 6-1

a mix of scheduled and emergency patient arrivals, data could be taken from the existing system which would be typical services times at any of the four blocks in Figure 6-1. These data could also be generated by a simulation based on tossing a pair of dice. Each toss would represent the passing of a small amount of tme, and a patient arrival during the time interval would be designated by the appearance of some combination, say 5, which occurs with a probability of 4/36. The arrival rate could be adjusted based on experience, and would therefore change to account for both peak and slack loads. Departures from the system would be simulated in the same way. Implementing this basic simulation strategy on a computer would be much faster and would utilize

1. A set of programs generating patient arrivals and departures.
2. A means of keeping track of the state of the clinic at all times.
3. An analysis of the system properties of interest.

In our example, a simulation study showed that the new organization significantly reduced patient congestion with the screening physician, registrar, and cashier, but revealed new tie-ups in the admissions office. These results provided a basis for an adjustment to a new design. An important aspect of this study of a clinic is the fact that the clinic staff was involved in the actual planning and data collection. The retraining and adjustments that were eventually required of the staff were better understood this way. Many examples of the simulation of elements of hospital and clinical systems exist in the literature, now including the operation of clinical labora-

tories, maternity wards, a teaching outpatient clinic, scheduling systems, and bed-allocation policies. All these simulation models required a rapport between the systems researcher and the personnel of the real-world system.

Community Health Systems

A system as complicated as a community health system requires not one, but many models in order to analyze and understand the operation of the overall system. A particularly useful way to distinguish the subsystems is by noting the extent to which the input resources are variables or parameters (3). From the total community resources, some portion is allocated to the community health system, and these resources are utilized at three levels, designated strategic, operational, and tactical.

The lowest level is the tactical, and here the health decision maker chooses from among the resources at hand. For example, the surgeon in a hospital must function with the resources provided by the hospital which are immediately available. His decisions are based on the needs of a specific patient, and any model used at this level must describe individual patient behavior.

At the operational level the health decision-maker's function is to organize available resources for the immediate future. The laboratory chief must decide how to meet the future demands on his laboratory with the personnel available. The statistics on which planning is based at the operational level are derived from numbers generated at the tactical level. In other words, the scheduling of the use of x-ray facilities must be based not only on the needs of the individual patient but also on system limitations and capabilities. Many models framed at the operational level exist. The literature includes operational models of queues in the doctor's office, menu planning, and various aspects of system management. These models have been based largely on existing mathematics and there has been little contribution to theories and concepts.

At the highest level are the strategic models, and these describe such processes as the size and location of facilities to be constructed and the kind of services to be offered. Parameters from both the tactical and operational levels are variables at the strategic level, and the time frame is very broad.

A health-care delivery system may be thought of as possessing the three levels just described, and the systems modeler must decide where to direct his research. In fact, however, the term "health system" is still a highly abstract concept. At this relatively early stage of research, the chief value of the health system modeling effort is to emphasize and dramatize the present lack of system in the organization and distribution of health resources. Results are aimed frequently at the goal of improving connections among health organizations.

Community Health Service
System Simulation Model (4)

Reference (4) presents an interesting systems analysis of community health services and then simulates one part of the system, a maternal and infant care program.

Community health should be regarded as the totality of all the health states of all the individuals in a given community. Because a state of health is a dynamic concept, static models will not suffice. The analysis we will be examining utilizes three major conceptual categories: basic health needs which establish demands, which, in turn, require community health resources. Kennedy defines a health need as the deviation of an individual's condition from certain prescribed health standards. Such standards are far from absolute. They depend on the state of current knowledge, the actual substandard condition of the individual, and the medical environment in which the evaluation takes place. Health needs may be established in three ways: actual, community-perceived, and individual-perceived. Actual health needs include those generated by inherited characteristics, socioeconomic conditions, and noxious agents. Community-perceived health needs are influenced by factors such as availability of trained medical personnel, methods, and equipment.

Accompanying each demand is a need, although the relationship may be obscure. A function of the health system is to determine which needs require treatment.

The demand portion of the illness cycle is the least understood. For example, what the individual perceives to be his health problems may be significantly different from the medical symptoms. A summary of the analysis of the conversion of a health need into a demand for health service must inclue some consideration of the following factors:

1. Economic: income, hospitalization plans, and

conceptions of cost associated with treatment.

2. Spatial: location of the necessary medical services and facilities for the patient.
3. Sociological: family size, religion, education, and ethnic background.
4. Psychological: the patient's current state of well-being and his attitudes and beliefs about illness, medical personnel, and medical services.

The health demands are satisfied by community resources such as manpower, facilities, materials, and techniques. Our objective in analysis is to develop a generalized health service system, and the subsystems have been identified as follows:

1. Promotion: the dispensation of all types of health information to the community.
2. Prevention: the reduction of the needs for health services by reducing the incidence of either disease or accidents.
3. Diagnosis: the determination of a treatment procedure based on an understanding of the symptoms, the disease, and the patient.
4. Treatment: the performance of the specified treatment with available resources.
5. Rehabilitation: the process that includes return to the well population, or death of the individual, or the definition of new health demands.
6. Health resource planning: the means by which the previous items in this list are planned and funded.

The several factors, entities, and processes we have discussed thus far can be brought together into a generalized health service system as shown in Figure 6-3.

Figure 6-3 represents symbolically a very complex reality. The question now is whether or not useful models can be developed from this abstract representation, and whether sufficient data will be available to support the modeling effort. Following this, we must verify the model and ensure that it has the quality of flexibility so that it can be applied in the health planning of other communities.

To demonstrate the needs, demands, and resources approach to simulating a health-care delivery system, a particular program was chosen. The simulation of a maternal and infant care program generated needs represented by pregnancies on the basis of population characteristics of size, race, and associated pregnancy rates. The probability that an eligible pregnancy will attend the clinic was the means of simulating demand.

For the maternal and infant care program, resource utilization is primarily of a preventive nature. The intent is to ensure a proper environment for the mother during pregnancy and for the child during the first year of life. The simulation was not designed to indicate optimal operation but to determine the effect of internal and external changes on the operation of the project. In this way the health planner is assisted in his selection of alternative programs for various conditions.

The model inputs were in the form of characteristics of

1. Population—size and pregnancy rate.
2. Pregnancy—outcome distribution.

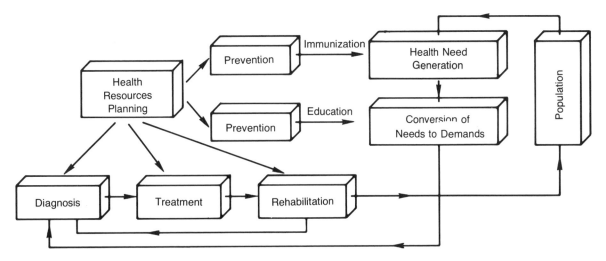

Figure 6-3

3. Mother—age distribution, marital status, education, income, and so on.
4. Clinic resources—number, type, and cost.
5. Infant illness—where treated, prognosis, outcome.

The model operations can

1. Calculate the number of monthly pregnancies.
2. Build a schedule of clinic visits for the mother.
3. Allocate resources for clinical schedules.
4. Establish pregnancy and infant records.
5. Generate an illness pattern for the infant.
6. Build a schedule of office and clinic visits for the mother and child.
7. Record resource utilization and costs.

The model outputs generate

1. Maternal clinic utilization reports, including case load, resources utilized, clinic operating time, and costs.
2. Infant clinic utilization report, similar to 1.
3. Yearly summary report of the clinic operations, including effectiveness measures.
4. Pregnancy termination report, which reports the outcome of all pregnancies terminated during the year.

Continuing the information listing necessary for a complex situation model, next we itemize the assumptions necessary for model development:

1. Pregnancies can be predicted by a crude pregnancy rate multiplied by an appropriate class size. From this, need can be determined.

2. Demand is a function of a constant level of community acceptance of the clinic and of the spread of information about the clinic. Information spread increases with time to an upper limit.
3. No micromodel of clinic dynamics is to be attempted. Resource utilization will be treated in an aggregative manner.
4. Clinic service time will be approximated by a truncated exponential function.
5. Clinic caseload and time available for clinic personnel–patient communication determine patient attitude toward the clinic.
6. Because of insufficient data, infant mortality rates are taken to be stationary, and the emphasis is on reduction of infant abnormality and infant morbidity.

The yearly summary report includes effectiveness measures, and those adopted for the model were

1. The number of scheduled visits divided by the number of actual visits, and this measure is designated the maternal clinic attendance ratio.
2. The total personnel utilization ratio, which is an effectiveness measure based on the total hours allocated. A ratio greater than 1 is an indicator of the overuse of personnel, less individual attention, and poorer patient attitudes. When these factors improve for a ratio less than 1, there is an accompanying increased cost per patient treated.
3. The abnormality ratio is an effectiveness measure defined as the number of project abnormalities divided by the project popula-

tion and multiplied by the reciprocal of the national abnormality rate. In this way, project abnormalities were calibrated against national levels.
4. The infant illness ratio compared project data to national averages. The number of project infant illnesses was divided by the project infant population and multiplied by the reciprocal of the equivalent population illness rate for the nation.

The heart of the model was four equations developed by regression analysis from project data. The first equation allowed the prediction of the number of mother's visits to the prenatal clinic:

$$
\begin{aligned}
\text{no. weekly visits to prenatal clinic} = {} & -(0.224 \times \text{no. weeks pregnant at first visit}) \\
& + (0.3547 \times \text{length of gestation at birth, weeks}) \\
& + (0.0002 \times \text{annual income}) - 2.427.
\end{aligned}
$$

The second equation predicted the total number of home and extra service visits:

$$
\begin{aligned}
\text{total no. visits} = {} & + (0.2816 \times \text{mother's age}) \\
& + (2.0374 \times \text{no. visits to prenatal clinic}) \\
& + (5.094 \times \text{complications of pregnancy}) \\
& - 7.7204.
\end{aligned}
$$

The third equation predicts the birth weight of an infant:

birth weight $= +$ (0.07738 × no. nutrition group conferences attended by patient)
$+$ (0.10112 × length of gestation at birth)
$+$ (0.3699 × parity)
$+$ 1.7624,

where parity equals 1 for the first child and 2 for a subsequent child.

The last regression equation allows the prediction of the number of infant illnesses:

no. infant illnesses $= +$ (0.0156 × no. weeks pregnant at first visit)
$-$ (0.677 × length of gestation at birth, weeks)
$+$ (0.101 × birth weight of infant)
$-$ (0.0001 × annual income)
$+$ (0.0149 × total number of home and extra service visits)
$+$ 2.155.

The data base for the simulation model we have been describing is substantial. The major data source was the mother and child records from the maternal and infant care program. Data from other sources included community population, number and type of clinic resources, and so on. Eight computer routines form the simulation model and process the input data to produce the four reports after model execution. The clinic reports list caseload, case characteristics, resource levels, resources used, and allocated costs. The yearly report includes measures of effectiveness, statistics based on monthly reports, and other miscellaneous data. Data on both the community population and project patients is contained in the yearly pregnancy-termination report. The model results match well with actual data.

The major point to be remembered about this particular simulation is that it possesses a basic framework that can be used to describe the community as an entity seeking solutions to health problems. The health needs of the community are converted to demands for financing, personnel, and facilities from the resources of the community, the state, or the federal government. Clearly, then, the simulation will assist in community health planning. An initial difficulty in implementing the model is the lack of understanding of various community processes. No theory of community health exists, and setting the objectives of community health and perceiving health needs is very difficult.

A health need is converted into particular actions at several levels, from individuals to the total community acting through a designated party. This process, however, is very inefficient because of limited information and the lack of an integrated approach. Consequently, difficulties that are incurred include duplication of resources, ineffective location of facilities, and inefficient use of personnel. Community health service system demands are generally higher than can be met; therefore, careful planning to eliminate the listed difficulties is important. The maternal and infant care simulation model that we reviewed demonstrates the usefulness of the application of the systems approach to a community health setting. The model is a unique alternative means for evaluation to assist the planner in adjusting the means to achieve the desired objectives. For example, the planner could manipulate the model to determine how the following items related to the input data:

1. The effects of variations in the pregnancy rate on clinic caseloads.
2. The effects of variation in the absolute size or in the migration rates for white and nonwhite populations on the project operation.
3. The effects of changes in staff levels on program effectiveness.
4. The effects of changes in the socioacceptibility of the clinic on clinic costs.
5. The sensitivity of
   a. Distribution of initial visits.
   b. Distribution of pregnancy outcomes.
   c. Distribution of mother's education.
   d. Length of gestation.

For the planner to deal with these listed eventualities without the use of the simulation model would be extremely difficult. Health-care delivery subsystems with an adequate data base are amenable to the techniques of systems analysis.

Community Hospital Design

In this section, for the first time, we will use the systems approach as an aid to design—and, in this case, the design of a community hospital. Current problems in medical care would be eased by new hospitals which were designed to place greater emphasis on ambulatory care rather than inpatient care, which avoided dupli-cation of facilities, and which provided more efficient patient care (5).

The shift to ambulatory care represents a growing change in philosophy which suggests that our medical resources would be better applied to keeping well people well, as opposed to making sick people well. The community emphasis, then, would be on preventive medicine.

Avoidance of the duplication of hospital facilities has been avoided in some communities by the use of legislative action. Individual hospitals have no mechanism to control their own growth. The consequence is expensive, low-utilization, duplicate facilities, usually in the more esoteric treatment rooms and equipment rather than in areas where payoff to the public might be greater. Examples are day care for the elderly or community mental health facilities.

Recent examples of attempts to make hospitals more efficient include the tightening of Medicare standards, frequent reviews of utilization, and management studies. Efforts of this sort have allowed the lowering of room rates.

The task of determining the optimal size and arrangements of the hospital organization can be accomplished with systems analysis and simulation. The results of the systems approach will be extremely useful to the hospital architect, the critical decision maker in this case.

A useful and appropriate objective of a hospital system may be defined as minimizing illness in a community, or the counterstatement of maximizing health in a community. This objective easily includes both in- and out-patient care. A possible approach after making this definition would be to measure the community illness, determine a reasonable standard for illness, and then identify the extent to which specified facilities would reduce illness to the community standard. Architectural decisions could be made to design the facilities to meet this standard at a minimal cost.

The approach just described sounds logical, but would be extremely difficult to accomplish; in addition, the focus on illness reduction would eliminate many other services from the hospital facility. For example, a hospital designed solely for illness reduction is not likely to have dayrooms, bookmobiles, social activities for the patients, a chapel, a physician's lounge, facilities for volunteers, classrooms for pediatric patients, or rooms for parents to stay with their children. The policy for discharging patients would be based on when they were well enough, regardless of the environment to which they would be returning.

Again, we have encountered the difficulty of defining objectives for a complex system. It appears that the contemporary community hospital serves many functions in addition to reducing illness. An invalid approach to hospital design used in the past was the specification of the required number of beds, followed by determining all other space and program needs from this.

Conway proposes that the design focus be on programs and assume that the programs will lead to the goals of the hospital. Using this approach the following steps would be appropriate:

1. Determine the range of medical programs which contribute to comprehensive medical care in the community.
2. Survey the community to determine which programs are being carried out elsewhere, or which are unacceptable in the hospital being planned.

3. The programs to be implemented must have their most effective size and arrangement specified. This step requires the use of a variety of mathematical models.

The range of programs that the new hospital design must accommodate are aimed at the goal of comprehensive care. The designer will include a basic set of programs which are the natural purview of the person or agency who has initiated the new design. Beyond that, other facilities that provide segments of care, such as physicians' offices, private laboratories, dentists' offices, pharmacies, public health facilities, neighborhood health centers, and mental health facilities, must be surveyed to avoid duplication and provide coordination. The community acceptability of each program must be considered. To do all this, a framework must be developed within which a broad spectrum of programs will fit. A suitable framework would recognize that hospitals provide medical care, education, research, and a social environment. Each of these items is a suitable program category. A matrix of programs, ranging from prevention to rehabilitation, is made possible by identifying types of patients and stages of care and combining them.

The hospital will be used for the following purposes:

	Patient Types	Means	For
Stages of care or illness			
Clinic visits	Maternity	Treatment	Patients
Physicians' office visits	Acute medical	Personnel	Personnel

	Patient Types	Means	For
Outpatient department visits	Dental	Equipment	Primary relations Volunteers and clergy
Purchase of drugs	Orthopedic		
Home care	Surgical		
Nurse visits	Pediatric		
Social service	Chronic		
Hospital admission	Mentally ill		
Hospital preparation for treatment	Tubercular		
	Patients requiring special treatment		
Treatment			
Care following treatment			
Discharge			
Rehabilitation			
Extended care			
Nursing home care			
Death			

Education and research

Social–psychological demands
  Social and religious
  Relief of anxiety
  Relief of boredom

Physiological demands
  Eating
  Dressing
  Toilet
  Waiting
  Sleeping

The preceding list of program possibilities will be reduced and adjusted according to unique local conditions. For example, some hospitals have extensive research programs, others have none. The choice for this option will depend on the attitudes of decision makers and community funding policies. A probabilistic model would be used to design the maternity department because births occur randomly and the admission dates are uncontrollable. In the case of a surgery program, demands would be forecast, then schedules designed to minimize the number of operating rooms needed, at some inconvenience to the surgeons. Tabulating hospital records of utilization would be used for gross decisions, such as intensive-care beds, the number of operating rooms, physical therapy treatment tables, and so on.

An interesting example of a dynamic model for planning patient care in hospitals was given by Wong and Au in reference (6). Their simulation model plans health services in a proposed hospital by simulating the predicted service demands on the basis of the health-care needs of the community and the hospital administrative policies for providing services. The dynamic nature of the model is provided by the fact that time-dependent data and features can be simulated in the facilities planning process, thus enabling admissions and service rates to be adjusted over various simulated time periods. A preliminary plan is developed according to the forecast service demands, which were generated by artificial random observations based on health-care statistics. When this plan is modified according to various policies for personnel and facilities, the most significant factors affecting the efficiency of health care in a hospital can be considered prior to its construction.

The preliminary plan just discussed evolves around providing the several services necessary to establish the critical levels required to maintain the current minimum health standards in the community. The flowchart given in Figure 6-4 depicts the process of conceiving a preliminary plan for patient care in a new hospital.

As an example of the detail implicit in Figure 6-4, the types of diseases included singly or in groups in the case distribution array are (1) infective and parasitic diseases; (2) allergic diseases; (3) diseases of the respiratory system; (4) diseases of the cardiovascular system; (5) diseases of the blood and spleen; (6) diseases of the digestive system; (7) endocrine, metabolic, and nutritional diseases; (8) diseases of the nervous system and emotional disturbance; (9) urogenital and venereal diseases; (10) diseases of the eye, ear, nose, and throat; (11) diseases of the locomotion system; (12) injuries and adverse effects of external causes; (13) obstetrics; and (14) special diagnostic cases.

The artificial observations generate the distribution of hospital cases by utilizing pseudo random numbers which represent appropriate distributions of the frequency of occurrence of various types of diseases.

The facility items which the simulation includes are (1) nursing units; (2) operating rooms; (3) delivery rooms; (4) medical, chemistry, and histology laboratories; and (5) electrocardiogram, radiology, and physical therapy.

The limitation of the planning model in Figure 6-4 is that there is no means of adjusting admissions and service rates through feedback paths

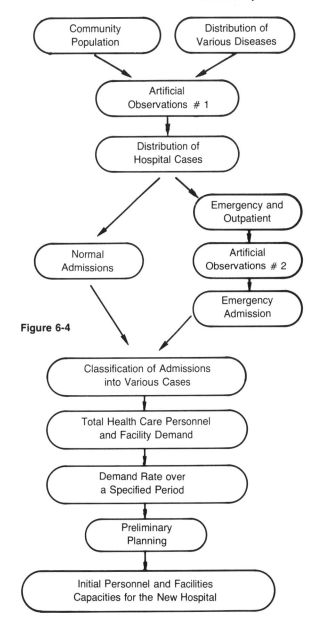

Figure 6-4

to match supply and demand. For example, the priority basis of admissions influences the waiting list, which in turn affects decisions regarding the length of stay for certain patients and the availability of beds. The dynamic features of Figure 6-5 include the following additional inputs:

1. The priority rating for admission.
2. The modes of variation in health-care performance.
3. The preliminary plan resulting from the previous stage.

The feedback path from the service performance block adjusts the admissions to meet predicted demands. The random generation of the daily total number of patients reflects seasonal variations and other time-dependent effects on certain diseases. Epidemics and natural disasters can also be injected as random factors. Admission control will allow the regulation of personnel and facility capacities when they are exceeded for short periods of time. The adjustment of the service rate of the proposed hospital is accomplished by changing service units or the capacities of personnel and facilities. When the service rate has been adjusted to meet the predicted demand, the final plan will generate detailed information for case treatments as well as personnel and facilities. If the simulation model is restricted by fixing hospital personnel and facility capabilities, then the model output is in the form of modifications and rescheduling in order to improve the efficiency of operation. It should be clear from these observations of the dynamic model that the planners and designers can use the model to investigate many pertinent factors affecting the design and operation of the proposed hospital.

A number of computer programs were developed to accomplish the many operations depicted in the two flow graphs. A few examples are listed below. Subroutines are able to

1. Change the demands for certain cases over a short period of time according to a particular mode of case variation.
2. Place certain elective cases on the waiting list according to some prescribed priority when the predicted demand exceeds service capacity over a certain period of time.
3. Generate health-care personnel and facility daily demands for all cases over any specified period.
4. Break down hospital cases into surgical, medical, and obstetrics cases according to the case percentages based upon statistical records.
5. Generate occurrences of elective, emergency, and out-patient hospital cases based on statistical records and utilizing a pseudo-random-number generator.

Figures 6-4 and 6-5 represent the two-stage operation of planning the health care of a proposed hospital. First, a preliminary plan is generated, and then the operation of such a plan is simulated over a long period of time. In this way, both the short- and long-term effects of proposed alternatives can be carefully evaluated. The alternatives may be loosely grouped as environmental factors and administrative policy decisions. Environmental factors include the growth or aging of the population, the rise and decline

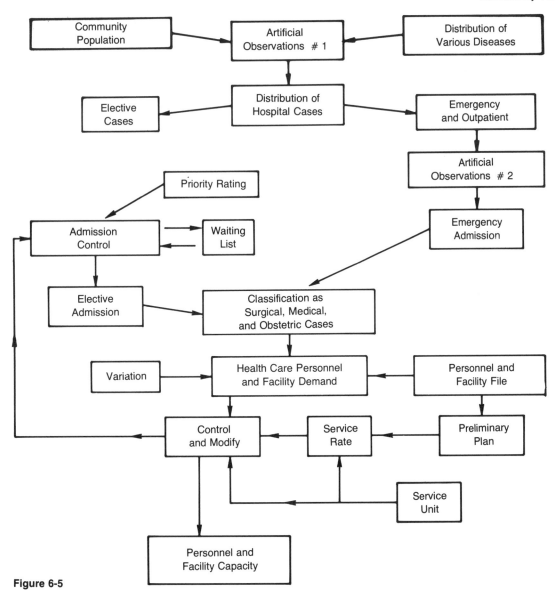

Figure 6-5

of certain disease rates, and trends of occupational hazards. Examples of administrative policies are the effect on the demand rate of the change of the average number of days of hospitalization for any type of disease, as prescribed by new medical practices. Another case is the tightening of restrictions on the admission policy for senile patients because of the increasing prevalence and availability of home health care. The change in demand for health services because of third-party payments requires the implementation of administrative policies.

The identification and inclusion of all significant influencial factors allows the simulation of a planning cycle for time ranges of up to 5 years. To plan operations in an existing hospital requires about 1 month.

In this paragraph we will review an example given by Wong and Au to demonstrate the planning of health care in a proposed hospital. We assume that a hospital is being planned to serve a community with a population of 250,000. Using the statistics, which give the annual averages of daily case distributions of various types of diseases per 10,000 in the categories of elective, out-patient, and emergency cases, the daily distribution of hospital cases per 100,000 population can be generated. This is done with the random-number generator, and the same means is used to generate the number of emergency admissions. At the beginning of the simulation the hospital is assumed empty and it takes about 14 days to stabilize the number of admission cases, which are automatically classified into medical, surgical, and obstetric cases. The treatment of each type of disease appears in the form of a demand file of facility and health care personnel. For example, for a surgical case with 7 days of hospitalization, the first 5 of the 15 items of the facility and health-care personnel demand file appear as follows:

1. General bed (beds):   1
2. Obstetric bed (beds):   0
3. Operating room (hours):   3
4. Delivery room (hours):   0
5. General laboratory (hours): 5

Use of the demand file allows the demand rates for services over a specified period of time to be accumulated. The service units for the 15 items of personnel and facility are specified by numbers corresponding to appropriate units given by the following:

1 1 9 20 8 8 8 12 12 8 8 8 8 8 8

The first two numbers have the unit of beds, the next eight numbers have the unit of usage hours, and the last five numbers have the unit of hours per shift per person. The required capacities of the same 15 items are listed in the preliminary plan on the basis of the average demand rate for all hospital cases and the peak demand rates for emergency and critical cases.

The dynamic model will be used to check the adequacy of the preliminary plan. For example, seasonal averages of daily case distributions of various diseases will give different results in a 30-day simulation than if annual averages are used.

The control and modify function will check the accumulated demand rates against the service rates. The criterion used by the simulation is that if the predicted accumulative demand rate in the next 14 days for any item of service exceeds the latter by 10 percent or more, it will cause the adjustment of the service units or the change of capacities in the preliminary plan when the adjustment of service units is restricted up to certain allowable ranges.

The hospital planning procedure we have been reviewing may be implemented in a variety of ways. Data representative of the particular community should always be used. The probabilistic distribution description used for the case distributions of diseases should be adjusted for the magnitude of the number of cases—normal distribution for a large number and Poisson distribution for the small numbers associated with rarer diseases. The number of beds required will be influenced by the prevalent trend of admission control by physicians who have no common criteria for admitting patients to the hospital. If the concern is space allocation, personnel and facilities capacities can be converted. Any information available regarding the prediction of community growth and other future events should be considered. If the model is being used to study an existing hospital, then real data obtained in the scene should be used. Ultimately, the most value received from the simulation is garnered because the planner can experiment with various alternatives on the simulation model to help avoid paying the price of making erroneous decisions in the real world.

An aspect of hospital design not yet discussed is the issue of arranging and relating space. Obviously, the designer has many alternatives. The criteria for design are many, and space arrangements will not be made for any single purpose, for example to reduce walking distance.

The following list is a beginning to establishing objectives to use in making space decisions. The possible objectives are efficiency, maximization of responsibilities, various social–psychological ends, and flexibility.

Efficiency	Examples
1. Minimize the expenditure of time	1. Out-patients to laboratory
2. Reduce the labor of walking	2. Outpatients to laboratory
3. Seek help quickly	3. Patient to nurse
4. Reduce the labor of carrying supplies	4. Nurse to supply rooms
5. Allow a logical sequence of transactions	5. Laundry operations

To maximize responsibilities	
1. Observe those transactions for which one is responsible	1. Surgical supervisor to surgery entrance
2. Control pilferage	2. Dietitian to supplies

For social–psychological purposes	
1. Allow clarity of route	1. Outpatient to laboratory
2. Comfort of travel	2. Patient to X-ray
3. Meet expectations	3. Patient to lounge
4. Provide reassurance	4. Patient to spouse
5. Reduce boredom	5. Patient to patient activities
6. Educate	6. Resident to surgeon
7. Encourage interaction	7. Specialists to others

To use spaces flexibly

Adjoin bedrooms which may be used by several patient types

Another factor to be considered in relating spaces are any environmental problems. Several environmental outputs should be minimized. An incomplete set of examples is given in the following list.

Output	*Examples*
1. Noise	1. Labor room, shop
2. Dust	2. Shop
3. View of the dead	3. Holding room
4. View of suffering	4. Treatment room
5. Privacy	5. Dressing room
6. Odors	6. Soiled-linen room
7. Radiation	7. X-ray room
8. Electrical interference	8. Elevator
9. Vibrations	9. Laundry
10. Congestion	10. Corridors
11. Potential explosions	11. Anesthesia room

An important issue not yet considered is the geographical location of the proposed hospital. The U.S. Public Health Service suggests several factors to be taken into account. The new hospital should be near the center of population and presumably then will be near the center of physicians' offices. Hospital sites should avoid mosquito-breed areas, industry, airports, heavy traffic, and a wide variety of nuisances and objectionable institutions. These criteria are basic; however, they need substantial expansion of concept in order to interpret. For example, more important than locating the hospital near the center of population would be a location where time in transit is relatively easy, where patients are in familiar surroundings, where their expectations are met, and where patient boredom is reduced.

The outline of hospital design information presented accounts for a complex set of variables and provides the basis for a crude model. A last category of architectural constraints requires some discussion. A tradeoff between functional requirements and aesthetic values must be balanced. The design for the hospital which the architect develops is itself a compromise, and some of the systems designer's objectives may be compromised in the process. Conway gives the following example of architectural constraints.

In an old 350-bed hospital there was always a waiting list for admission, and administrators assumed that either an addition or a new hospital was needed. A review of records revealed, however, that in spite of waiting lists, there was always some space in some nursing units. It appeared, then, that a new patient-reassignment procedure and a different space-arrangement policy would solve the problem. The plan was to adjoin nursing units with similar types of patients and in this way make it possible for each unit to expand or contract according to need. The head nurses were interviewed to determine the compatibility of patients. Medical records were studied to detect seasonal variations in patient populations. In an attempt to model the mathematical distribution of patient census for each nursing unit, it became apparent that there were a number of uncontrollable and unpredictable variables due to changes in hospital policy. For example, the initiation of Medicare was followed by a 12 percent increase in the patient census of the medical nursing units. However, the most serious constraint was architectural. The multistoried design of the hospital limited the alternative arrangements among the 11 nursing units to very few. The depth of the study made it possible to solve the waiting-list problem by the design of a reassignment procedure and by the new method of space arrangement.

This example exemplifies the problem of architectural constraints which exists in both new and existing buildings. The interface between systems techniques and architectural implementation suggests that upper stories should not be inordinately larger than lower stories; hilly and sloping sites limit solutions, prevailing wind direction must be considered; and aesthetics must be maximized.

Community Planning for Public Health

New York City has attempted to incorporate systems analysis techniques in its public health planning. As early as 1966 a document entitled "Systems Analysis and Planning for Public Health Care in the City of New York—An Initial Study" was published (7). Topics touched upon were the structure and roles of municipal health-care organizations; health needs of the divergent communities within the city; environmental health variables; roles of community clinics, hospitals, nursing homes, philanthropic, and private health agencies; requirements for ambulatory, extended, and chronic care; trends in private practice, group practice, and prepayment services; and the impact of federal health legislation.

The use of systems methodology in the community health setting dealt with several classes of problems. The first issue was concerned with the basic documents which specify the requirements for health practices. These documents establish codes which formalize the operation and responsibilities of city agencies and tend to render them highly resistant to change. Changes occur as additions rather than deletions or recombinations, and the isolation and autonomy is further reinforced. An additional complication is caused by federal and state funding agencies, which support certain health programs and exclude others. This kind of rigidity is reflected at the organization level. The autonomy of function of the departments of health, hospitals, welfare, and mental health of the City of New York is a case in point. Each has its own programs, budgets, facilities, and service, and any advantages of this arrangement are outweighed by the difficulties of a broad program integration. Even though a common population is served, the divided authority has led to duplication of very costly services, fractionation, separate administrative staffs and the like. Coordination is difficult and must come from a high level, in this case, the office of the Mayor.

Delivery problems represent a second class of problems. The planning and implementation of low-cost, high-quality medical care for everyone has not been realized. High cost of care exists because of a morass of entitlement laws, prepayment plans, group-practice methods, and reimbursement formulas, and the results delivered have been extremely variable. Efficient satisfaction of human medical need is everywhere thwarted by chaotic conditions, duplication of services, and episodic treatment.

Another aspect of the delivery problem has been created by each of the autonomous health agencies establishing separate districting at the community level. The effect is to cause poor coordination between agencies and to defeat continuity of care when the health problem of an individual cuts across jurisdictions. Medical records are a part of this problem. Since good patient records are only rarely established, and even more rarely retrievable, the basis for any diagnosis may be the current examination only, without reference to visits at other clinics or hospitals of other agencies in other districts. An individual may have medical records at several points in the health-care delivery system, and it appears that the total aggregation of the medical record is not possible. A proposal for an organization to solve some of these problems is given in Figure 6-6. In Figure 6-6 Public Health deals with disease prevention, environmental health, community research, and central laboratories. Clinics and Ambulatory Care are responsible – for hospital-based, community-related diagnosis, treatment, and referral. Hospital Administration provides for health-care-delivery-system unification and improvement. Chronic Care Division oversees all rehabilitation, nursing homes, home care, and various programs for the aged.

Similar classes of problems exist in the areas of environmental health. Separate agencies exist to deal with housing, welfare, sanitation, pollution control, and so on. Communications and coordination between separate agencies have been inadequate. The air-pollution aspect of community health cuts across many levels of organizations and agencies.

These several classes of community health problems we have reviewed are being attacked, usually independently; their interdependence is not recognized frequently. The laws have forced the creation of overlapping agencies, which have created overlapping boundaries, which, in turn, reinforce poor medical delivery. The most fundamental fact regarding medical care and environmental health, which we have emphasized in previous chapters and which is repeated here, is that none of these factors can be considered

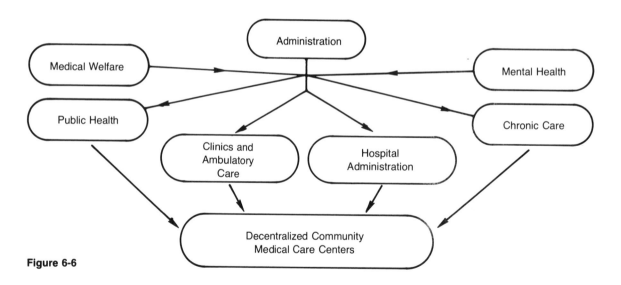

Figure 6-6

singly if they are to be overcome. Our continuing point is that the systems approach is a promising methodology to allow simultaneous consideration of all pertinent factors.

Community Health and Environmental Hazards

In the previous section there was brief mention of environmental health issues. In this section we will examine how community health issues arising from environmental hazards affect population redistribution and in turn influence all community planning (8). As communities increase in population, significant portions of land area are converted to urban use. This, in turn, leads to the growth of environmental hazards, including land, air, and water pollution; noise; brush fires; floods; and earth slides. The county of Los Angeles is a prime candidate to review some of the resulting effects. In 1960 there were more than 6 million people in the 4083 square miles of Los Angeles County. Topography includes mountain ranges, inland valleys, deserts, and coastal regions. Governmental and commercial agencies provide census tract distributions of the "artificial" health hazards of smog and airplane noise, and the natural or economic hazards of brush fires, floods, and earth slides. Statistical analysis showed that gross population densities are related directly to the presence of artificial hazards and, with the exception of flood, inversely to the presence of the natural hazards. In addition, the data show that with the exception of flood, population densities for each natural hazard combination are less in areas subject to two natural hazards than in areas with no or one natural hazard, and artificial hazard areas have higher population density than areas with or without natural hazards. Other data analysis showed that percentages of land area with artificial and natural hazards are, respectively, directly and inversely related to settlement maturation.

The data discussed thus far has shown that urbanization is associated with the contemporary pattern of environmental hazards. It remains to be shown that hazards play a role in activating population growth away from affected sites. In the Los Angeles area, data on the smog hazard can demonstrate this very well. When the date of settlement is controlled, a larger share of the 1950–1960 decreases in smog exposure of population and dwelling units can be accounted for by movement to areas that are smog-free than to other factors.

The environmental hazards we have been discussing, along with other conditions associated with a high degree of settlement maturation, represent an overutilization of some urban land. Subsequent population changes then favor hazard-free areas. If there are general propositions implied by these results for Los Angeles County, then there is a need for demographers to examine artificial and natural hazards of microurban environments with respect to their influence on urban demographic and ecological phenomena.

Community Waste Disposal

To attempt to consider carefully the effect of land, air, or water pollution on community health would require at least one book on each type of pollution. In this section we will take a brief look at how the systems approach can solve problems in community waste management (9).

The community waste problem can be stated simply. If as much material is not taken out of a community as is brought in, eventually something desperate will happen. The complex and sophisticated transportation networks discussed in an earlier chapter have no counterpart to take material out of the community. The removal of waste appears to be almost a community afterthought. Until recent times the natural sinks of land, air, and water and nature's destructive processes have made it possible to keep the sight and smell of waste under control with a little effort from man. Now, in spite of the fact that cities spend several billion dollars annually, a solid-waste crisis is at hand, and the frightening implications of the impact on community health are part of it. A curious counterpart of the urban problem is that the agricultural community also has a waste-removal problem. Both the urban and agricultural settings, however, are subject to the same problems: shortage of disposal sites, changes of processes and materials, and a renewed strong concern over the health hazards of some of the traditional modes of waste disposal. And the force behind all these problems is the growing population.

The technology of waste handling is in primitive condition. The 1965 Solid Waste Disposal Act was a federal attempt to give new life to the field of solid-waste research. The availability of federal research funds has been continued in air-pollution legislation because solid-waste disposal has been recognized as a prime source of air pollutants. With the growth of population and the accompanying growth of per capita waste disposal, technology alone is not the answer. Pertinent federal, state, county, and municipal agencies will have to cooperate in new ways. New systems of incentives may have to be legislated to encourage an economics of disposal to complement the usual economics of production, supply, and demand. New kinds of organizations may have to be established to plan, capitalize, and manage new types of disposal systems. A public education program is needed to provide information to individuals and organizations so that they can be encouraged to adjust their own mode and style of living to reduce waste.

Current estimates are that 6 to 8 pounds of waste material are generated per person per day. And this figure is double the weight of 40 years ago. Predictions are that the per capita waste will double in 20 years. Most of this increase is due to the incredible growth in the use of paper products and packaging of various kinds. A simple example is the great mountain of waste paper associated with buying a meal for a family at a drive-in restaurant. The volume of paper waste is increasing even more rapidly than the weight because of a trend toward thinner grades of paper that pound for pound take up for more space when crumpled as refuse.

Most people in this country still have their trash collected by trucks and taken outside the city limits. Sixty percent of the refuse from the nine counties surrounding lovely San Francisco Bay was dumped along the bay shore in the recent past. Open dump burning was common until the more sophisticated techniques of the sanitary landfill and municipal incineration. The idea of the sanitary landfill is to cover over each day's production of refuse with a layer of dirt to contain the odors and emerging fly pupae and to exclude rats and moisture. Clearly this process takes a lot of space and dirt—an acre of ground piled 7 feet high for every 10,000 people every year. New York City has very limited landfill sites. In the meantime, it costs nearly $30 per ton to collect, transport, and dispose of the refuse in a land-fill site, and that is about three times the cost of a ton of West Virginia coal, mined and delivered in New York. High-temperature municipal incinerators are designed to reduce the total volume of wastes by 75 to 90 percent, thereby cutting the demand for land fill. Unfortunately, incinerators are notorious air pollutors. Although incinerators can be improved to meet air-pollution standards, both capital and operating costs are extremely high—five or six times as much as the cost of sanitary land fill.

Many of the wastes that nature is being asked to deal with she simply cannot degrade. Glass, aluminum, plastics, and detergents are well-known examples. Many toxic chemicals are released in air, water, and on land by industrial processes. Economic factors continue to doom hopes of recycling waste materials, especially those which are not degradable. One hopeful note on recycling is the suggestion that certain wastes be stored and saved until economic and technical factors are more favorable.

Perhaps the most frustrating aspect of the problems of solid-waste disposal is that the issues are not glamorous enough for our leaders and governmental units. Citizens appear not to react unless they are directly affected. It seems that the culture requires a crisis condition before

any real commitment to solving the problems will occur. It is not uncommon to have individual towns spend large sums on their own land-filling equipment—and it stands idle most of the day. The same town continues to pollute their downstream or down-wind neighbor with sewage discharge or air pollution. The control of air pollution frequently causes the diversion of the pollutants to rivers. Control of water pollution often pollutes the land, and the control of solid-waste disposal usually pollutes all three—land, air, and water.

In Chapter 2 we discussed cost-benefit analysis. Applying that technique to the case of pollution is extremely difficult. Aside from evaluation, it is hard even to list all the direct and indirect costs to society of air, water, and land pollution. How could we put a price on an eyesore, or the twice-daily change of shirts, the long commuter trip to a cleaner suburb, the foot cut on a beer-can tab, the physical ailments that appear related to environment hazards, or the lack of a sense of well-being of a citizen? Even if these evaluations could be made, would the data persuade city dwellers to consider waste disposal when purchases are made, and to dispose of their wastes in such a way as to lessen the long-term drain on agricultural or industrial resources? Probably not!

The most promising sign of potential solutions to these problems occurs in university settings, where the air-, land-, and water-pollution experts are drawing together into "environmental engineering" groups where "waste management" is the byword and the systems approach is the problem-solving methodology. In addition, in these areas of the county where even minimal efforts have been made toward cooperation among political jurisdictions, the results have been impressive. In Los Angeles, collection and disposal costs—even with high prices and long hauls—are among the lowest in the country because of the sharing of land-fill sites and economies of scale.

The systems approach suggests that regardless of the method of disposal, a regional approach must be used. The region could, for example, include several communities, but they should be formed as geoeconomic entities rather than along political boundary lines. Cost-benefit studies conducted under these conditions could prevent economic overkill. When all possible pollutants have been identified, their sources and relative disagreeableness known, then the money spent can be used to the greatest advantage. Some experts believe that communities are proposing to make rivers and streams cleaner than is economically justified, and that the money might be better spent on solid-waste or air-pollution problems.

Assuming no substantial innovative approach, the major cost in any solid-waste disposal, whether conducted on a regional basis or not, will be collection and transportation. By conventional techniques 75 to 90 percent of municipal refuse expenditures are required for the fleet of trucks with their supporting personnel that make the collections from individual households. An unattractive by-product is the nuisance that garbage trucks make in the form of noise, odor, and traffic congestion. Investigators have conceived a variety of schemes to deal with the collection and transportation problem. One possibility is to use computers to keep track of the random movement of railroad cars, and to load the empty cars exiting cities with bales of refuse. In this way longer hauls would be possible to more remote locations for land-fill operations. Europe already has scattered locations which use tubes to pump the wastes out of town. The waste is moved pneumatically to control incinerators as far as 1½ miles away. Another version of the pumping scene would grind the refuse, mix it with water, then pump the slurry out of town. Experiments have shown that one pipe 2 inches in diameter could easily carry the wastes of a town of 10,000 to 15,000. Such a waste system could be coupled with the existing sewer system and, again, pumping to remote locations is possible.

The schemes just mentioned are likely to be expensive to establish; however operating costs would be low. The primary ingredient required is some sort of utility company organized specifically to plan, build, and operate such a regional network. The impetus to establish such an organization is likely to come only from a crisis in spite of the fact that present approaches to waste handling are inadequate, expensive, and wasteful of natural resources.

References

(*1*) W. J. Horvath, The Systems Approach to the National Health Problem, *Management Sci.*, **12**(10), B-391–B-395 (1966).

(*2*) C. D. Flagle, The Role of Simulation in the Health Services, *Am. J. Public Health,* **60**(12), 2386–2394 (1970).

(*3*) D. Howland, Toward a Community Health System Model, Chap. 9 in *Systems and Medical Care*, A. Sheldon et al., eds., The MIT Press, Cambridge, Mass., 1970.

(*4*) F. D. Kennedy, Development of a Community Health Service Simulation Model, *IEEE Trans. Systems Sci. Cybernetics,* **SSC-5**(3), 199–207, 1969.

(*5*) R. H. Conway, Operations Research and Hospital Design, presented to the 40th National ORSA Meeting, Oct. 28, 1971.

(*6*) A. Wong and T. Au, A Dynamic Model for Planning Patient Care in Hospitals, *IEEE Trans. Systems, Man, Cybernetics,* **SMC-2**(2), (1972.

(*7*) H. M. Adelman, System Analysis and Planning for Public Health Care in the City of New York, *Arch. Environ. Health,* **16** (Feb. 1968).

(*8*) M D. Van Arsdol, Jr., Metropolitan Growth and Environmental Hazards: An Illustrative Case, *Ekistics,* 48–50 (1966).

(*9*) T. Alexander, Where Will We Put All That Garbage, *Fortune* (Oct. 1967).

CHAPTER 7

JUSTICE SYSTEMS

The complete justice system comprising the agencies of police, prosecution, courts, and corrections needs to be examined in a total systematic way. A number of problems exist that can be approached only if the issues are examined in an integrated way. Major crime summaries made by Federal Bureau of Investigation show that crime rates for all major categories are increasing dramatically. As an example, robbery rates per 100,000 individuals increased 70 percent in the period 1960–1967, and assault rates increased 51 percent in the same period. The State of California appears hardest hit, with an FBI total crime index of 3207.5 crimes per 100,000 individuals compared to a national average of 1921.7. Studies are being conducted to determine the critical factors from among such possibilities as rapidly increasing population and social and economic conditions.

In this chapter we will emphasize criminal justice systems and the urban components of prosecution, courts, and to a limited extent, corrections. Police activities are examined in Chapter 8. Both criminal and civil courts have experienced severe problems of congestion and delay for many years. In 1849 the New York Commissioners on Pleading and practice reported (1): "It is well-known that in New York City the New York State Supreme Court is weighed down by the accumulation of former years. . . . Unless relieved of that load, it can never perform its proper functions in respect to accruing business."

As an example of the magnitude of the chronic problem of delay, the following figures were derived from a survey of adult arrest cases which entered the Manhattan branch of the New York City Criminal Court in 1969. The typical misdemeanor case required approximately 4 months for processing; 5 percent of such cases lasted 1 year or longer; felony cases took even longer.

Criminal Justice System

The criminal justice system (CJS) has remained remarkably unchanged through the significant social, technological, and managerial changes of recent decades (2). The CJS is well insulated from other institutions and has relative freedom from external examination and influence. In addition, the individual components of a CJS operate with relative independence, and nowhere is there a single administrator of a CJS with control over all the constituent parts. These several unusual conditions are being examined by various agencies in an attempt to solve some of the problems of the CJS, and it is from this effort that has come the need for models that will permit the study of a total CJS. Currently there is almost no knowledge of the cause-and-effect relations in a CJS; however, these first-generation models will, as a minimum, identify the data needs and the research questions. The models will be used also for resource allocation and for examining the effects on crime of actions taken by the CJS.

A highly simplified version of the CJS is given in Figure 7-1. Crimes originate in society from individuals who have (recidivist) or have not previously committed a criminal act. Many of these crimes are not detected or are not reported. Only a small fraction of crimes committed result in the arrest of a suspect. Of all the suspects arrested, again only a small fraction ultimately receive a

sentence from a judge as one of the following forms:

1. A monetary fine.
2. Probation, usually with a suspended sentence.
3. Probation, following a fairly short jail term.
4. Assignment to a state youth authority.
5. A short jail term (usually less than 1 year).

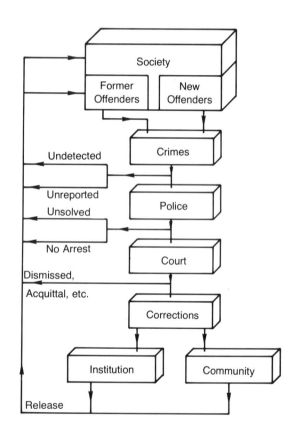

Figure 7-1

6. A prison term normally no less than 1 year at a state institution.
7. Civil commitment for some specified treatment.

The CJS of Figure 7-1 also processes probation and parole violators and juveniles. Juveniles are treated far more informally with much more freedom of choice exercised by the juvenile authorities.

The simplified flow graph of a CJS given in Figure 7-1 and the cursory description which follows suggest two approaches to modeling the CJS. The first possibility is the *linear model,* in which the emphasis is the flow through the system and the accumulation of costs flowing from a single arrest. This model lists the workload, personnel requirements, and costs associated with types of crimes and projects all these planning variables as functions of future arrest rates. The second modeling approach is a *feedback model* which considers the recidivism probability associated with each released defendent and his subsequent processing for future arrests.

Linear Model of a Criminal Justice System

The costs, workloads, and manpower requirements at the various stages of the CJS are essential aspects of the linear model of a CJS. The workload is defined as the annual demand for service at the various processing stages, for example courtroom hours or detective man-hours. The *manpower* requirement may be computed from the workload by dividing by the annual working time per resource. The total operating costs are allocated to offenders by standard cost-accounting procedures. The flow of people through each processing stage of the CJS will be represented by a vector with several dimensions. A useful set of flow dimensions which can be utilized are the seven index crimes which the FBI annually tabulates. These are willful homicide, forcible rape, aggravated assault, robbery, burglary, larceny of $50 or over, and auto theft. Thus, the total flow through the CJS will be characterized in terms of these seven measures of types of crimes. The input to the model is the number of crimes reported to the police during 1 year, and the model outputs are the computed flows, costs, and manpower requirements that would result if the input and the system were in steady state. At each processing stage the input flow is divided into the appropriate output flows by mutiplying the input vector by the probabilities of each of the output alternatives. A simple example of the jury-trial stage processing is given in Figure 7-2. The figure depicts the jury-trial stage for a particular crime which is one dimension of the input vector. The definitions of the symbols of the figure are as follows:

N = number of defendants who receive jury trials,
N_g = number of jury trial defendants found guilty,
N_n = number of jury trial defendants not found guilty,
P_g = probability that a jury trial defendant is found guilty.

P_g is called branching probability, and there are seven different components of P_g required as input data for this stage.

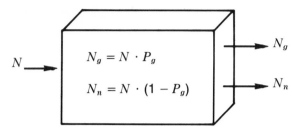

Figure 7-2

A somewhat more complicated example, which will give more detail, is the prosecution and courts model given in Figure 7-3. The input to the model is the vector, N_T, the number of adult arrestees who are formally charged with one of the seven index crimes. There are seven output vectors based on the seven sentence types. Figure 7-3 shows four intermediate output vectors which identify people being processed by the CJS who never reached the sentencing stage. These are

1. N_s = number of people formally charged who do not reach trial stage.

2. N_d = number of defendants whose cases are dismissed or placed off calendar at the trial stage.
3. N_{n1} = number of jury-trial defendants not found guilty.
4. N_{n2} = number of bench- and transcript-trial defendants not found guilty.

Four types of branching probabilities are required by the model in Figure 7-3.

1. Whether or not the defendant reaches the trial stage.
2. The type of trial or whether dismissed at trial stage.
3. The trial verdict.
4. The sentencing decision.

Each of these probabilities will be set by a *careful examination of data*.

Using these probabilities the flow through each stage of the CJS can be calculated, and then the costs can be computed as the product of the unit costs and the flow rates. The processing stage

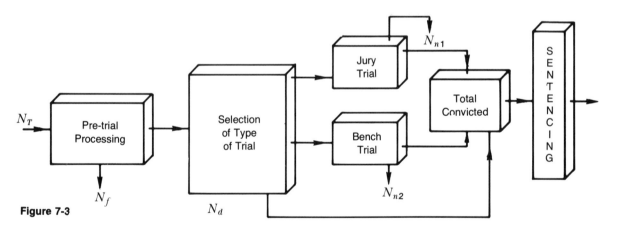

Figure 7-3

flows allow calculating annual workloads in terms of total trial days for both jury and judge trials and man-days for pretrial detention in jail. Using these results, the annual manpower requirements, such as prosecutors, judges, and jurors, can be calculated on the basis of unit productivity such as annual trial days available per prosecutor.

The linear model we have been discussing is very complex, but it barely scratches the surface of the reality of a CJS. Any given processing stage is, in fact, composed of a number of detailed processing stages in the real system. To have included more detail would have exceeded the data capability and added complexity to the model with little additional payoff. The simplifying linear relationship between flow and cost implies that all costs are variables; however, many costs are fixed and independent of flows. Obvious examples are the costs of fixed facilities such as courthouses. This is a steady-state model, and variables are assumed to be constant and independent of each other and of exogeneous variables. Examples of variables that violate the assumption are detention times and the probability of a prison sentence. In both cases, the variables are likely to be a function of demand.

In spite of these drawbacks the linear model of a CJS does permit a reasonable first estimate of costs, workloads, and flows, and allocation of these to crime type and processing stage. In addition, under specified conditions, the linear model can be used for projections.

At previous points in the book we have mentioned sensitivity analysis. The linear model of a CJS can be used for this analysis to determine the effects of changes in one subsystem on the workload, costs, and manpower requirements of another subsystem. For example, the utilization of an improved fingerprint-detection system would increase the number of arrests per burglary, which would, in turn, require planning for the increased cost and workload effect on the subsequent court and corrections subsystems. Two quantities must be defined in order to discuss sensitivity analysis.

Given any two system flows C_i and N_i we define $\delta C_i/\delta N_i$ as the incremental change in C_i per unit change in N_i and $(\delta C_i/\delta N_i)/(C_i/N_i)$ as the incremental fractional change in C_i per unit fractional change in N_i. As an example of the use of these two definitions, assume that C_i designates the cost at some stage of processing people charged with crime i. If N_i represents the flow of persons at an earlier stage and C_i is linearly related to C_i in the form of the equation

$$C_i = A_i + B_i N_i,$$

then

$$\delta C_i/\delta N_i = B_i = \text{average additional cost}$$
incurred for processing at the later stage per additional person charged with crime i and injected in the earlier stage of the CJS,

$$(\delta C_i/\delta N_i)(C_i/N_i) = B_i N_i/(A_i + B_i N_i) = \text{average}$$
fractional increase in cost incurred at the later stage for processing people charged with crime i per unit fractional increase in individuals charged with crime i and inserted at the earlier stage.

In other words, the first definition is an *incremental cost per person,* and the second is the *fractional increase in cost per unit fractional increase in the number of people.*

As a more specific example of $\delta C_i / \delta N_i$, let C_i be the number of jury trials for robbery defendants and N_i be the number of adults charged with robbery in magistrate's court; then the incremental number of robbery jury trials per additional robbery defendant from magistrate's court is calculated to be 0.1. An alternative interpretation of this value is that 0.1 is the probability that a randomly selected robbery defendant from magistrate's court will proceed to and have a jury trial.

The linear model of a CJS has also been used to estimate future requirements. Projections of future workloads, costs, and manpower requirements are valuable to CJS administrators at all levels. Projections are made by investigating the constancy of branching probabilities, and by linear extrapolation. Blumstein and Larson reported projections on the number of arrests per reported crime, final disposition of adult felony arrests, future numbers of crimes and arrests, and increases in the values of CJS variables.

Like all models, continued use will improve its quality. Other useful analysis considered for future application are

1. The effect of the introduction of new police hardware, such as automated fingerprint files.
2. The effect of more widespread provision of free defense counsel on prosecution and court workloads.
3. The greater use of nonadjudicative treatment, such as the use of social service agencies as an alternative to prosecution.

4. A change in sentencing policies and how it might affect the development of new correctional facilities or the hiring and training of additional parole and probation personnel.

Feedback Model of a Criminal Justice System

The feedback model of a CJS to be discussed in this section takes a very different viewpoint from the previous linear model. The function of the model is to follow an individual criminal as he cycles and recycles through the CJS. Applications of the feedback model include:

1. Given an individual's age and crime at first arrest, the model can compute the expected crimes for which he will be arrested at each age.
2. Using the cost results of the linear model, the model can calculate the average cost to the CJS of the criminal career of an individual.
3. The effect on cost and criminal activities can be determined as rearrest parameters are varied.
4. The model provides a unified theoretical framework within which to study the process of recidivism, and a structure on which to test proposed CJS policies on recidivism.

A general flow graph of the feedback model is given in Figure 7-4. Flows are identified by crime type as before, and each flow variable designates the offender's age. The number of arrests per year by crime type and age of individual at first arrest plus the recidivist arrests constitute the input to the model. These total arrests are then processed through the feedback model. The feedback element of the model computes the number of people, their crimes and date of rearrests, who recycle back through the CJS after dismissal or release. The characterization of an offender at each possible dismissal point is based on age, prior criminal record, and rearrest probability. The number in the flow through the CJS is multiplied by the rearrest probability to determine how many will be rearrested. The distribution of delay between release and the next arrest is used to calculate the age at rearrest. As a last step, the crime type of the next arrest is found from the crime switch matrix.

The flows in the feedback model can be interpreted in two ways. The first possibility is as a cohort-tracing model, where a career of that cohort can be traced. The second use of the model is for population simulation, where the input is the total present distribution by crime type and age of first arrests, then the computed flows represent the steady-state distribution of all people, including recidivists processed by the CJS.

Four branching probabilities are required in the feedback model of Figure 7-4:

1. The probability that an arrested adult is formally charged with a felony.
2. The probability that an adult who is charged will be placed in a state correctional institution.
3. The probability that a person who is charged will be placed on probation or in a local jail.
4. The probability that an adult who is charged is dismissed before or during trial.

As in the linear model, *all the values of these probabilities are established from existing crime statistics.*

Perhaps the most intriguing aspect of the feedback model are the details associated with the re-

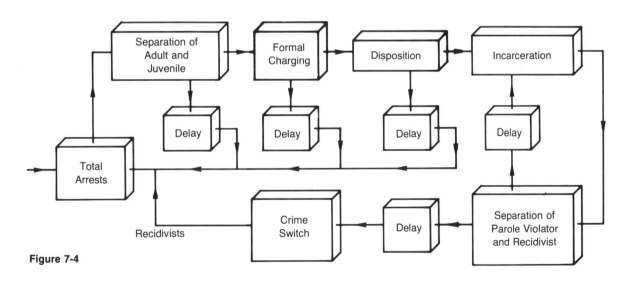

Figure 7-4

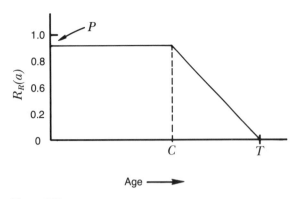

Figure 7-5

arrest probabilities. Rearrest probabilities are defined as a function of age and crime at the last arrest, and the feedback model requires that rearrest probabilities be specified at each point of dismissal from the CJS. Examination of existing crime statistics indicates that the variation of rearrest probability usually follows a gradual decrease after the age of 30 years. Figure 7-5 graphically portrays the variation of rearrest probability.

The symbols of Figure 7-5 are defined as

P = probability of rearrest of people who are less than C years of age at the time of release,

C = age at which rearrest probability starts decreasing linearly to zero,

T = age beyond which arrest does not occur.

Some examples of parameter values for the rearrest probability function under the condition that the offender was formally charged but not found guilty are

1. Homicide: $P = 0.65$, $C = 40$, $T = 100$.

2. Robbery: $P = 0.80$, $C = 35$, $T = 80$.

Under the condition that the offender is found guilty and either placed on probation or in a local jail, the parameters change as follows:

1. Homicide: $P = 0.25$, $C = 35$, $T = 100$.
2. Robbery: $P = 0.573$, $C = 30$, $T = 80$.

These sample data and other data indicate a substantial decrease in the likelihood of rearrest for those placed on probation, even though they were found guilty. We offer this as an observation, not as a general proposition; however, the observation indicates how components of a theory can grow from use of the feedback model.

The several delay blocks in Figure 7-4 account for the time between release from some point in the CJS and rearrest. Data describing these times are very rare, and the time distributions were chosen to have a mean of about two years based on existing data. A typical distribution function relating time delay and rearrest probabilities would set the rearrest probability at 0.5 for the first year, 0.3 for the second year, 0.1 for the third year, and nearly zero for all later years.

All recidivists pass through the crime switch of the feedback model. The function of the switch is to compute the crime type of the next arrest from a rearrest crime switch matrix where the matrix element P_{ij} is the conditional probability that the next arrest is for crime type j, given that rearrest occurs and the previous arrest was for crime type i. For the seven index crimes, 42 independent probability estimates are required to specify the matrix. Again, adequate recidivist data to estimate these probabilities is rare. Recidivist studies indicate that there is a strong tendency for offenders to be rearrested for crimes different from the previous arrest.

The general difficulty of limited area data throughout the model has necessitated a number of simplifying assumptions.

1. The age of the offender, the crime type of the last arrest, and the disposition of the last arrest are the only dimensions used to predict future criminal behavior.
2. The future crime of the recidivist as generated by the crime switch matrix depends only on the crime type of the last arrest.
3. CJS branching ratios are not a function of age or previous criminal activity.
4. Delay times until rearrest are based only on the disposition of the previous case.

In spite of these assumptions, the model has produced a number of useful results. First and most important, the model has identified significant data of the CJS, and once suitable record keeping is established, provides the framework in which to use the data. A number of runs of criminal populations have been processed by the model, and at this early stage of model development, the authors characterize the results as "modest." Another valuable use of the model has been to investigate the reduction of recidivism probability and its effect on criminal careers.

It is important to remember that the goal of both the linear and the feedback model has been to describe the operation of the CJS, and to assess some of the effects of this system on future criminal behavior. The achievement of these goals permits answers to questions regarding the costs, workloads, resource requirements, and rehabilitative procedures of the CJS. The deterrent effects of the CJS and public and private means outside the CJS used to control criminal behavior are not considered.

Justice Systems Data

We have already emphasized that data from the CJS has a variety of difficulties associated with it. A recent report by J. B. Jennings states that the data problems underscore the need for a major overhaul of the entire information system serving the criminal justice agencies in New York City (3). The data difficulties he cited are of four types:

1. Data reported by the various criminal justice agencies are incompatible with each other.
2. The data contained in many reports pertain to an aggregation of numerous physically distinct operations.
3. There are not uniform procedures throughout the CJS.
4. Long delays are invariably associated with receiving statistical reports of a CJS.

Each of these problem categories is worth expanding. An example of the incompatibility of statistics is the definition of the basic unit being counted: the Police Department counts people, the Criminal Court counts its docket numbers (there may be several numbers for one defendant), the grand juries and the Supreme Court count indictment numbers, and the city and state departments of correction count people. In addition, many commonly used terms are defined differently by different agencies. Example terms are defendant, youth, and reporting period.

The aggregation of data is such that except by examining the individual case papers, it is not possible to determine the exact stage at which the charges against "discharged" defendants were actually dismissed. Court reports are set up in such a way that there is no breakdown of the stage at which charges were modified or of the precise modification.

The lack of procedural uniformity is exemplified by the different procedural treatments of defendants by the district attorney's office and the grand jury.

The typical delay on reporting crime statistics is 2 years. This substantially reduces their usefulness.

Quantitative Models of Criminal Courts (4)

In a previous section we reviewed models designed to deal with aspects of the entire CJS. In this section we want to look at models that are directed to understanding and eliminating congestion and delay in criminal courts. In the past, responsibility for dealing with these problems has rested primarily with judges and lawyers. Recently the systems approach has been contributing to the effective administration of criminal courts in two ways. Models have been developed to deal with the allocation of court and related resources (Blumstein and Lavson's model contained the court as part of the entire CJS) and the scheduling of case appearances.

A detailed simulation of court operations would be based on a model which relates the flow of cases to the court operations that give rise to such flow. For example, we could develop a model for the court system shown in Figure 7-6 by treating each box as a queue with a specified processing time distribution and availability schedule, by specifying transit times on each branch in the diagram, and by indicating the rate at which cases are disposed of at each stage. An early example of this approach is that of Navarro and Taylor (5).

Micromodeling of court operations has also been done (6). The model simulated the court day minute by minute, taking into account the details of each case to be processed and the actual activities of each judge, district attorney, and defense lawyer. The model operates by calling cases from a calendar and processing them according to activities that require time blocks. The model has been used to compare court operations under alternative scheduling procedures and under a variety of allocations of judges, district attorneys, and public defenders.

As with all models, the intention is to offer the court administrator an opportunity to test alternative procedures and resource allocations without disturbing the ongoing operations of the court itself.

The second interesting class of court models is the scheduling models. Then function is to decide how to select a date in the future on which a particular case will make its next court appearance. The number of cases scheduled to appear in a particular court segment on a specified day is set in advance, and each case is scheduled to the first acceptable unfilled day in the segment in which the case is to appear next. The size of the court calendar must be determined. A suitable strategy to determine the calendar size is to bal-

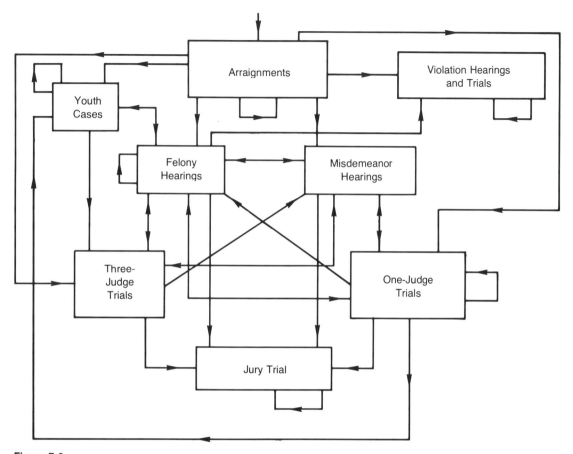

Figure 7-6

ance the waste of court time due to underscheduling against the waste of the time of other participants whose cases must be adjourned with action due to overscheduling. On the basis of observing court activity over a period of time, the relationship between the average underscheduling, the average overscheduling, and the number of cases in a particular court segment may be estimated as in Figure 7-7. X designates the number of cases calendared per day and from Figure 7-7 $(X_3 > X_2 > X_1)$. Using this information for a particular judge and court segment and considering the relative costs of overscheduling and underscheduling, the most desirable calendar size can be selected.

The application we are discussing is computer-

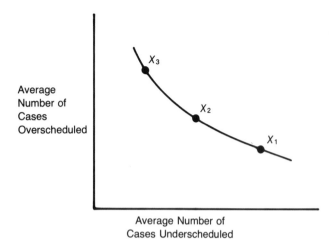

Figure 7-7

controlled scheduling of court activities. The computer programming will be designed based on the results of the simulation model. One cause of congested courts is that judges tend to select future dates primarily on the basis of the length of time until the next appearance of the available participants. Thus the court calendar becomes a function of adjournment policy because the judge has no up-to-date information on the size of future calendars in all court parts. The result is a variable-calendar-size technique of court scheduling as shown in Figure 7-8. The calendar of the hearing and trial parts (HT part) is formed as shown in Figure 7-8 and must accommodate both arraignment and HT part adjournments. To develop a scheduling model around these ideas requires the following processing assumptions:

1. Cases cannot be distinguished from each other as far as dispositions and adjournments are concerned.

2. A fixed fraction of cases not disposed of at arraignment is adjourned to each HT part.
3. Of those cases appearing on the calendar of any HT part on any day, a deterministic fraction is disposed of, and the remaining adjourned cases have an additional appearance scheduled.
4. Of those cases adjourned in any HT part, a fixed fraction is adjourned back to the same part and the remainder is divided among the other HT parts.
5. No consideration is given to the number of cases previously scheduled, selecting a date on which an adjourned case is next scheduled. Instead, a fixed fraction is scheduled to appear on the following a day, a fixed fraction on the second day, and so on.

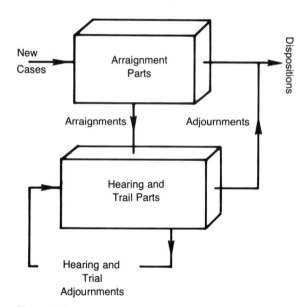

Figure 7-8

From the previous observations and assumptions, a scheduling model can be developed in the form of a closed-form mathematical expression for the number of cases that will appear on the calendar in every HT part, on every future day, in terms of the number of new cases received per day and the various processing parameters. To give specific details of the model, we will examine the case in which all HT parts are identical and the system has been in operation for a long time.

Let
- c = size of the calendar in each HT part,
- n = number of cases adjourned each day from the arraignment parts to the HT parts,
- m = number of HT parts,
- f = fraction of cases on a day's calendar that are disposed of that day;

then we can show that the steady-state number of cases on the calendar in each HT part is given by

$$c = \frac{n/m}{f}. \qquad (7\text{-}1)$$

This equation is a mathematical characterization of the following verbal statement: The number of cases per day that will be scheduled to appear in each part is inversely proportional to the fraction of cases disposed of, and directly proportional to the number of new cases received in each part each day. There are four major possibilities for the relationship between the total number of cases disposed of and the calendar size. These cases are depicted in Figure 7-9. Case e represents the disposal of the entire calendar and case a the idealized case where a fixed fraction of

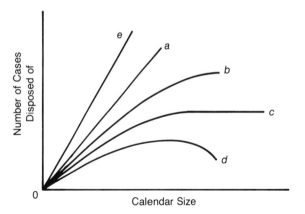

Figure 7-9

cases is always disposed of, regardless of the calendar size. The remaining cases portray the more practical situation where the fraction of cases that are disposed of decreases as the calendar size increases. In case b the number of cases disposed of continually increases, in case c it increases until a plateau, and in case d, the number first increases, then decreases.

To develop more of a feeling for this court scheduling model, let us explore case a. To do this we will refer to Figure 7-10. Our problem is a familiar one in other settings, although cast in the language of a CJS, it may seem obscure. We want to find the calendar size (c) and the fraction of cases on a day's calendar that are disposed of that day (f) for the conditions of case a. The conditions of case a which must be satisfied simultaneously are that cases c and f must be related by Eq. (7-1) and f_a must be a constant. The graphical solution of the two simultaneous equations is shown in Figure 7-10, where the constant

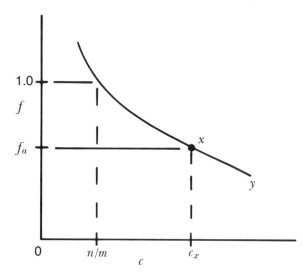

Figure 7-10

value of f_a is shown and where the curve y is a plot of Eq. (7-1). The point x identifies (c_x, f_a) as the values that simultaneously satisfy the two requirements. Therefore, the steady-state calendar size for case a is c_x, which can be computed by

$$c_x = \frac{n/m}{f_a}. \tag{7-2}$$

The other cases are more complex but can be dealt with by a similar approach.

Of course, more complex scheduling procedures may be designed into other simulations. Some of the additional scheduling intricacies which would be useful include a priority measure for cases, allocation techniques that save portions of calendars for high-priority cases, and a means by which participants in each case can inject their preferences for various dates.

Reference (7) has an interesting approach to the simulation of the court scheduling problem which emphasizes the efficient use of jurors. Many other quantitative aspects will be included in the future, and most existing models require additional testing and validation.

Corrections Simulation

Although the corrections system of the CJS is only partially an urban concern, in this section we will look at aspects of the corrections system for completeness. We will examine a methodology which uses a mathematical model to distribute the corrections population to each of four programs (prison, parole, probation, and jail) and then computes the total cost of operating each program (8). The portion of the CJS we will study is given in Figure 7-11.

The model that predicts the population in any correction program is

$$p(t) = Tp(t-1) + s(t), \tag{7-3}$$

where
$p(t)$ = column vector whose components are the corrections-program populations at time t,

T = transition matrix which accomplishes program reassignment, including release by utilizing matrix entries which are probabilities,

$s(t)$ = column vector whose components give the number of persons not currently under corrections supervision but committed to the program.

$$s(t) = ARc(t), \tag{7-4}$$

where

A = allocation matrix which reflects the court sentencing policy,

$c(t)$ = column vector whose entries are the number of free people who were convicted of charged offenses during the year,

R = transformation matrix whose components are the numbers of free persons in each prior-record category convicted for each charged-offense group.

Thus $s(t)$, the people who are convicted but not yet in a corrections program, is formed by modifying the convicted people by the court sentencing policy and by their prior record. The sentencing policy and prior record information act as weighting factors to transform $c(t)$ into $s(t)$. Rewriting the basic model equation, Eq. (7-3), in matrix form with words inserted as entries will help clarify the details of the information contained in the equation:

$$
\begin{bmatrix}
^P\text{prison} \\
^P\text{parole} \\
^P\text{probation} \\
^P\text{jail} \\
^P\text{termination}
\end{bmatrix}_t
= [T]
\begin{bmatrix}
^P\text{prison} \\
^P\text{parole} \\
^P\text{probation} \\
^P\text{jail} \\
^P\text{termination}
\end{bmatrix}_{t-1}
+
\begin{bmatrix}
^S\text{prison} \\
0 \\
^S\text{probation} \\
^S\text{jail} \\
0
\end{bmatrix}_t ,
$$

where

$$
[T] =
\begin{bmatrix}
^T\text{pris, pris} & ^T\text{pris, paro} & ^T\text{pris, prob} & 0 & 0 \\
^T\text{paro, pris} & ^T\text{paro, paro} & 0 & 0 & 0 \\
0 & 0 & ^T\text{prob, prob} & ^T\text{prob, jail} & 0 \\
0 & 0 & ^T\text{jail, prob} & 0 & 0 \\
^T\text{term, pris} & ^T\text{term, paro} & ^T\text{term, prob} & ^T\text{term, jail} & 1
\end{bmatrix}
$$

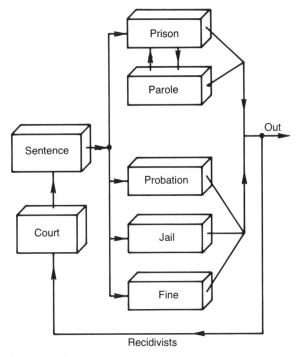

Figure 7-11

The theoretical foundation of the model in Eq. (7-3) is based on the identification of the corrections process as a Markov process. The process property which establishes this is the fact that the current population depends on the population of the previous time period but not on the population any earlier. The properties of the Markov chain allow the matrix (T) to be written in a form that permits the calculation of the mean number of years in each correctional state. The vector $s(t)$ is obtained by operating on the number of convictions during the year for each offense category considered:

and

$$pris = prison,\qquad prob = probation,$$
$$paro = parole,\qquad term = termination.$$

This simulation equation identifies the corrections population, $p(t)$, in any year as a function of both the population in the system during the previous year, $p(t-1)$, and the new sentences to the corrections program, $s(t)$, during that year. The transition probabilities matrix, T, describes the movement of the corrections population during the year. The data needed to begin to operate the simulation is $p(0)$, the population already in the system; T, the probable movement of the population during the year; and the new population $s(t)$, received by the system. $p(0)$, the populations in the four corrections programs (prison, parole, probation, and jail) is available in annual criminal statistics.

To determine the movement between correction programs in order to characterize T, we must look at the allowable transitions. These were implied in the word version of T and are given graphically in Figure 7-12.

Again working from criminal statistics, the probability of moving between correctional programs can be computed from actual population shifts in a given year. A sample T matrix with representative values is

	prison	parole	probation	jail	term-ination	
	0.679	0.238	0.045	0	0	prison
	0.307	0.422	0	0	0	parole
$[T] =$	0	0	0.652	0.09	0	probation
	0	0	0.081	0	0	jail
	0.014	0.340	0.222	0.91	1	termination

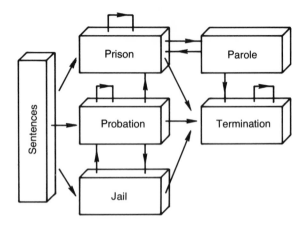

Figure 7-12

In order to form $s(t)$, we need data on $c(t)$, R, and A. The entries in $c(t)$ are the number of new convictions during the year for each of the 10 felony offense groups. Using felony data for recent years, and then curve fitting, future felony data can be predicted. Assuming a linear model, the percentages of felonies based on past data can be projected to future data. In this way $c(t)$ is formed.

In a previous discussion of R and A, we pointed out that each term of R represents the fraction of those convicted for an offense who have a prior criminal record, and the entries of A represent the fraction of those convicted for a particular record and offense group and who are sentenced to corrections programs. All these data are available in the court's records of felony complaint filings.

The effort to this point has been to generate the population of the corrections system in any given year. Data are available for the per capita system inmate expenditures for recent years and, using curve fitting, the equation relating prison costs to time was found to be

prison cost in dollars $= 2001.28 = 149.36(t + 7)$.

Using both projected populations and per capita cost, future program costs can be predicted. Our example has been for prison programs, and parole, probation, and jail costs are calculated in a similar way.

The simulation model contains court sentencing policy implicitly reflected in the rates of commitment of convicted defendants to the various correctional programs. The nature of the current offense and the existence of a prior record are the most important factors determining the sentencing policy. It is these factors which were used to construct alternative sentencing policies. Using the simulation, 5-year projections were made for four different sentence policy assumptions. The target sentencing policy refers to a lower commitment rate based on the experience of several counties with a lower commitment rate for each offense and record category. The four sentencing policies simulated were

1. Continuation of the current sentencing and present mean-time-served policy (the control case).
2. Continuation of current sentencing policy; reduction in mean time served.
3. Target sentencing policy; continuation of mean time served.
4. Target sentencing policy; reduction in mean time served.

The strategy of the four policies is based on combinations of high and low commitment rates and high and low mean time served in a correctional program. The important point is that the simulation model is set up to test alternative sentencing policies prior to their implementation. This particular study showed that the new sentencing policy was capable of affecting considerable savings for the correctional programs.

Crime-Prediction Modeling (9)

The capability to predict crime would be useful in at least two obvious ways. Clearly, the police could make many kinds of operations and planning decisions. In addition, a part of the population involved in the commission of crimes becomes the input to the CJS and therefore influences both resource allocation and scheduling. Very little research has been done on crime prediction; therefore, there was little precedence on which to make modeling decisions. The three levels of modeling decisions to be considered were (1) type of model and dependent variables to be predicted, (2) independent variables and functional relationships, and (3) parameter-estimation techniques. Some of the pertinent details of each level are listed next.

1. a. Types of models: extrapolative, associative, quasi-causal, causal, pattern recognition.
   b. Dependent variables to be predicted: number of offenses or arrests, probability of a crime event, specific crimes, calls for police service, and so on.
   c. Geographical boundaries: block, census tract or reporting district, division, city, county or region, state, nation.

d. Frequency and length of prediction: hourly, daily, weekly, monthly, quarterly, annually.

2. a. Independent variables: proxy variables, such as time; correlated variables, such as population; variables resulting from theory, such as measures of social pressure.

b. Functional relationships: linear or nonlinear, feedback control systems, and others.

3. a. Mathematical: regression, exponential smoothing, Fourier analysis, spectral analysis, moving averages, and others.

b. Nonmathematical: Trial and error, simulation, "eye-balling," physical modeling, and others.

c. Treatment of outlying points resulting from unusual occurrences and clerical errors.

Most of these modeling issues we have encountered previously, but to reinforce their importance we have repeated them as applied to the new area of crime prediction. What is especially important and interesting in this work is the means by which the basic model was chosen. The model choices and a brief explanation are given next.

1. Extrapolative model: Historical data are extrapolated into the immediate future by a variety of means.

2. Associative models: Objects that are associated in some way with the phenomenon of interest can be used to predict those phenomena. An example would be that if two divisions of a city were alike in an appropriate

sense, crime data from one division could be used to predict crime rates in the other division.

3. Quasi-causal models: The factors whose time series are highly correlated with the crime-time series of interest are identified. Predictions are based on the assumption that the observed correlations will continue and the expectation that the highly correlated variables are at least proxies for the real causes.

4. Causal models: If the factors that influence the variables of interest can be identified, and the ways these influences occur are either known or theorized, causal models can be constructed to express this knowledge or theory.

5. Pattern-recognition models: the intention of this type of model is to isolate the crimes committed by single individuals or gangs, and using this information to identify likely crime targets and likely suspects.

At the study site eight years of crime statistics were available, and it was decided to design a crime-prediction model based on the first 6 years of statistics, and then use the model to predict the existing next 2 years of statistics. In this way the model would be validated.

The 6 years of crime statistics were used to evaluate 54 models. The 54 models were formed by choosing three promising functional equation forms, three criteria for rejecting outlying data points, and six parameter-estimation techniques. The first 6 years of data were applied to each model, and prediction errors determined for the seventh and eighth years of data. The data were broken up according to the 27 crime types recorded and the best model fit to the data was determined for each crime type. In all there were 648 crime statistics time series for which models were chosen. In the study the results showed that the exponential model was chosen more than half the time, suggesting that crime was growing exponentially in the study area. Prediction uncertainties were about 15 percent.

The critical decision in the development of the simulation model was that predictions would be produced by extrapolation of historical time series. Mathematical equations used for such an extrapolation have to be capable of characterizing trends, seasonal variations, and cyclical variations. The choice of time as an independent variable assumes that the dependent variables do change with time. The rate of change of many processes is dependent upon the current state of that process, thus leading to exponential behavior with time. Seasonal variation was incorporated by the use of additive seasonal constants. Because of the modest data base, possible cyclical variations were not included in the model.

As we reported earlier in this chapter, crime data suffer from a number of problems. Reporting criteria differ substantially depending upon place and time. As an example, in 1950, when New York City introduced centralized crimes record keeping, the reported number of burglaries jumped 1300 percent over the preceding year. The geographical problems discussed previously influenced this study. The jurisdictional boundaries of police units in Los Angeles, the source of the crime statistics, underwent 6 major and 30 minor boundary shifts during the period 1958 through 1968.

Another data difficulty is that unusual events are commonly recorded in the standard categories. In Los Angeles during the Watts riot in August 1965, burglary arrests were 10 times larger than normal for August because of looting directly associated with the riots. The time series of crime statistics had to be adjusted for these unusual data. It is well known that a large fraction of crimes committed are never reported to the police. This fraction can change significantly because of several factors, such as a police–community relations program. The crime data must be adjusted for any change in the reporting fraction. The difficulties associated with the data, and the complex nature of the phenomenon being predicted, account for the complex strategy used to choose the form of the simulation model. This study and its systematic approach to modeling the process of crime prediction is an extremely interesting example of modeling methodology.

The uses of crime-prediction techniques are still largely speculation. Some of the possibilities are:

1. Long-range forecasts can be made for personnel recruiting, police academy curricula, police force structure, and equipment acquisition.
2. Alternative force structures can be evaluated; tactics and equipment effectiveness can be assessed.
3. Where, when, and how much of police tactical forces can be deployed will be more easily known.

The extrapolative model we have been discussing will give more precise short-term predictions as additional model development is done. Most experts anticipate that the particular activities of the police have an influence on the amount of crime, and therefore a general tool for assessing the effects of alternative actions will be of considerable value. However, any general tool must include many more dimensions than have been mentioned thus far. Social forces and processes must be characterized, and social conditions such as educational levels, unemployment, housing conditions, and so on would be part of the model. Further model evaluation is likely to develop feedback loops and be more causal in nature.

Urban Design and Crime Control (10)

Recently a provocative study of the relationship between urban design and crime control has been described by Angel (10). From his work it appears that environmental factors have a demonstrable and direct relevance to the matters of crime control and prevention. Angel's work is speculative and runs the risk of being indentified with single-purpose planning, but many of his ideas are worthy of our attention.

One of the first attempts to relate city planning and the discouragement of crime was made by Jacobs(11). Her position was that the evolution of the city into specialized activity areas has substantially reduced a major crime deterrent, that is, the voluntary surveillance of streets and public areas. She advocated an urban design model which distributed commercial activities along residential streets in order to increase surveillance. Unfortunately, in the evolution that has already occurred in many of our cities, little surveillance capability is present. For example, Oakland, California, has approximately 1200 miles of streets but only about 4 miles of frontage of all establishments that remain open in the evening.

Theories of crime prevention establish three basic categories: punitive prevention, corrective prevention, and environmental prevention. Primitive prevention uses threat of punishment to discourage crime; corrective prevention tries to eliminate the factors of motivation before they cause criminal behavior; environmental (mechanical) prevention places obstacles in the way of potential offenders so that it is difficult or impossible to accomplish the offense. Urban planning that has contributed to the reduction of overcrowding in housing, the creation of viable neighborhoods, the rehabilitation of slums, meaningful recreation programs, and community mental health clinics were, in part, controlling crime by corrective prevention.

To pursue the idea of environmental prevention requires focusing on the features of areas where criminal acts are committed, as contrasted with areas where criminals live. It is the physical elements in the city: buildings, fences, lights, and so on, and the technology of crime prevention: alarms, locks, closed-circuit TV, and so on that we must examine. The study assumes that there is not an excessive degree of rationality and planned behavior to criminals. Instead, the opportunistic unplanned element of finding a victim in circumstances where the criminal act can be carried out is the dominant mode. For example, locked cars are stolen less frequently than cars left with keys inside. The specific function of this opportunistic element in crime is not well studied or understood, and we have no cogent theory to draw on. In the absence of a theory we have no complete classification of a quantifiable set of determinants or of the relative weights that these determinants bear on the creation of criminal opportunities. Some of the possible factors can be classified into the categories of behavorial characteristics of the offenders and social deterrents of crime.

One behavioral characteristic of offenders is that they have a tendency to cling to areas of the city where they can function inconspicuously, feel secure, and where knowledge of the terrain assures fast and efficient escape. Another behavioral characteristic is the penchant for criminals to commit crimes within easy access of their place of residence. Thus, any area that does not have easy access to areas of criminal habitation will be less susceptible to the opportunistic element of crime.

The behavioral characteristics of people can increase the probability of their becoming victims of crimes. Whether a potential victim is alone, intoxicated, weak-appearing, loudly affluent, or highly distracted may contribute to their involvement in a criminal act. Data appear to indicate that these factors are more significant than the potential payoff on the location and timing of robberies and purse snatches.

Social deterrents to crime include police patrolling and the presence of police, community attitudes and awareness, the number of effective witnesses, and general visibility conditions.

The issue now is to consider possibilities for reducing the number and intensity of criminal opportunities through manipulation of the factors just reviewed which are capable of control. In a practical sense, this means control of the physical environment—the spatial configuration of the city and its buildings. Clearly, however, social,

political, and economic measures must be integrated with adjustment of physical features to assure an effective reduction of the opportunities for crime. In Angel's work this is done through a city that will maintain its safety through mutual surveillance of the citizens present on the scene. A critical hypothesis is the relation between the number of people present in an urban area and the number of crimes committed in the same area. Figure 7-13 graphically depicts the relationship. The figure shows a condition below the peak number of crimes where there are too few potential victims, and above the peak where there are too many witnesses present. The peak of the figure occurs in a range where there are enough potential victims present, but not enough people to provide for an adequate surveillance function. Most crimes take place in this critical intensity zone. The issue then becomes how to adjust the urban environment so that pedestrian circulation is always of such a level that the critical intensity zone is not invoked. To translate this requirement into physical terms is not an easy

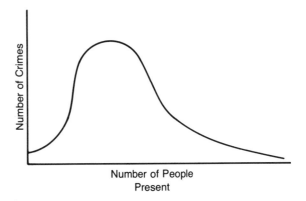

Figure 7-13

matter. However, Angel proposes 13 physical configurations for urban environments which will have the effect of discouraging crime. Because his work emphasizes only one dimension of the urban environment, crime prevention, it is likely that other welfare measures are being sacrificed. However, the point in his study, and our point in this section is that this poorly understood and usually neglected dimension of urban design should be considered along with all other design requirements and an optimal combination achieved by some means.

Angel's proposals include the centralization of evening establishments, the division of strip commercial development into sections which are deserted in the evening and those which remain open, and the development of "evening squares" where weather control is provided. The difficulties of implementation of most of the configurations would be substantial, especially in existing cities. These proposals would be considered simultaneously with proposals of other planners advocating other community interests. Much additional work needs to be done to validate the basic hypothesis, such as experimental data gathering, studying new housing developments with respect to the crime problem, and developing more sophisticated urban environment configurations.

Transaction Approach to Criminal Statistics (*12*)

Time and again we have returned to the point that CJS statistics currently available cannot provide a basis for any detailed analysis. If better reporting methods were available we could expect:

1. Better statistics to determine the impact of crime, to determine the effects of CJS policies and operation, to forecast the effects of changes.
2. Better allocation of resources to establish standards of performance, to identify areas where increased expenditures will bring maximum benefits, to adjust the use of basic criminal justice resources to social priorities.
3. To predict agency workloads in relation to both crime incidence and internal system factors.
4. Information for planning, evaluation, and daily decision making for legislators and administrators at all levels of government.

A review of the state of the art in crime statistics reveals that

1. Police statistics are the most highly developed reporting system.
2. Prosecution statistics are not available on a national, state, or local scale.
3. No national probation statistics exist.
4. National prisoner statistics are weakened by variations in the policies concerning kinds of sentences and offenders to be sent to the state and local institutions.

Many of these reporting incapacities are a direct result of our traditional concepts of the administration of justice. It has been traditional to give local agencies the responsibility for defining crime and developing a response to it. As a result, different approaches are exhibited in the variety of administrative structures and policies that translate penal code and criminal procedures into actions, and that allocate funds to what is viewed as a serious crime or a serious offender. The CJS is loosely structured and poorly defined. The subsystems often seem to possess contradictory goals, and this combination is ill-suited to develop comparable or consistent statistics on crime. Little attempt is made to reconcile the data output of different agencies.

An approach proposed by Wormeli and Kolodney is intended to resolve many of the difficulties just reviewed. It is called *offender-based transaction statistics* and it focuses on the individual arrestee/defendant offender and follows the processing of the individual from point of entry in the CJS to point of exit. The individual defendant is the only unit of count common to all criminal justice agencies and processes. An individual criminal record would not be unlike the individual medical history to which the health-care delivery system aspires. In this way the aggregated experience of someone who passes through the CJS will describe the functioning of the CJS.

The individual defendant is the unit of count; therefore, if a person were arrested for a second offense after being processed for a first offense, he would become a second-processed defendant. When several individuals are arrested for a single offense, each is regarded as a separate case.

These types of data can do anything the older systems can and, in addition, they can

1. Describe how the CJS operates to process defendants, and how agencies and functions relate to one another.
2. Determine processing time for individuals in the CJS.
3. Determine who the clients of the CJS are.
4. Grossly define much of the CJS in terms of

the number of subjects involved and the time span required per subject.
5. Provide a critical ingredient to the four dimensions of the basic data base: agency, offender, event, process.
6. Identify areas for which the investment of additional effort will be most productive.
7. Provide a basis for program evaluation.
8. Simulate changes in one part of the system and project the impact.
9. Be the basis of establishing normative statistics to detect the point at which the processes are out of whatever limited control is possible.
10. Be the basis of experimental research in criminal justice.
11. Have the ability to perform multidimensional analysis for selected groups such as offenders and crimes with the hope of uncovering causal mechanisms.

The status of research now in CJS is to make recommendations for a data-collection system to test any model against the actual environment.

Long-Range Planning in the Criminal Justice System (*13*)

A useful way of viewing long-range planning is that it is a systematic way of enriching the information base for decision making. The CJS has multiple decision-making groups, and long-range planning provides coordination to avoid overlap or omission among agencies or programs. In addition, the design and comparison of alternative actions for goal achievement is accomplished.

Realization of these two objectives of long-range planning can be thwarted by some of the reasons listed next.

1. The involved agencies resist any evaluation. All bureaucracies appear unwilling to divulge any information or data that will permit an external evaluation of their activities. For example, it is common in municipal governments for city agencies to fail to present alternative actions to the mayor for his choice. Instead, only the agency-preferred alternative is promoted adequately; other alternatives are either not divulged or are made to appear inferior.

2. Often there is a separation of planning and control, and if this is the case, it is very unlikely that any plan will be implemented. Tieing planning and budgeting processes together ensures implementation, and this is one aspect of the planning, programming, and budgeting methods discussed in Chapter 2.

3. Many planning staffs fail to involve high-level decision makers and thereby doom their plans to failure. It is not uncommon that many urban planning efforts are isolated from the mainstream of the political decision making processes.

4. Long-range plans must be based on the planner's conception of what the future will hold. A common error is the failure to build sufficient flexibility into any plans so that unforeseen future conditions can be accommodated.

5. Any long-range plans developed without *serious* consideration of *all* alternatives can expect major modification or rejection as time passes.

6. Unless an evaluation technique is built into a plan, there is no way to find out whether or not the plan is fulfilling its objectives.

To do long-range planning in the CJS requires consideration of these six dangers, but first we will summarize in this context the current state of affairs in the justice system today. The objectives of most CJS agencies are either ill-defined or inconsistent among agencies. Tremendous effort goes into dealing with day-to-day problems which occur because of historical conditions. In every part of the CJS there are more demands than capacity. As a result there is a shortage of resources everywhere. The quality of administration in most CJS agencies is very low. Paper work abounds, most processes are slow-moving, friction between participants is high, and problems are frequently referred rather than solved. Evaluation criteria throughout the CJS is very minimal, and there is little basis for judging the effectiveness or efficiency of individual agencies. Although this summary is pessimistic, many of the problems stem from the enforced fragmentation of efforts caused by the separation of functions.

After the preliminaries of generalized long-range planning and the state of the CJS, the following list consists of specific topics that require coordinated and strategic planning.

1. Agencies with overlapping jurisdiction in different problem areas must coordinate their activities.

2. The CJS is characterized by arrestees flowing through a number of agencies as their case progresses. Referral decisions should consider the quality of treatment or service that can be expected from each alternative agency to which referral can be made.

3. Any change in legal or administrative procedures and constraints can have a substantial effect, say in work load, on an agency further along in the process.
4. Research studies should be conducted in such a way that the results are generalizable and useful to all agencies affected.
5. Operations planning must occur for all areas that require coordination action, such as riot control, crime investigation, criminal intelligence, or criminal processing.
6. The type, number, and training of all personnel and their appropriate facilities requires long lead-time planning.
7. Equipment such as communications or data-processing systems require several years to acquire.
8. Reorganization must be planned to incorporate all the changes desired at that time.

The planning process outlined in this section requires a great deal of analysis. It is in this way that systems analysis and the systems approach can contribute significantly to long-range planning for the CJS.

Criminal Justice Laboratory *(14)*

Implicit in all that we have written on the CJS has been the need for basic research. So much of what happens in the CJS is based on expediency. In this final section we will look at an exciting plan for the development of a laboratory of criminal justice. Using funds appropriated under the auspices of the Omnibus Crime Control and Safe Streets Act of 1968, the California Council on Criminal Justice designated Ventura County as a criminal justice laboratory for the purpose of testing various configurations of the CJS. This occurred on August 26, 1970, and the implications of a laboratory county for research are enormous. The laboratory approach to problem solving in any social system is very difficult, and the concept of researching the actual social system in its realistic setting has seldom been invoked. To do it requires a basic reconsideration of the basic tenets of a justice system. The objectives of the laboratory county are given as:

1. Serve as a test setting for various configurations of the CJS.
2. Produce data on the relative effectiveness and economy of various programs of police, courts, and corrections.
3. Test hypotheses developed to describe phenomena of the CJS.
4. Test planning strategies and the associated data requirements.
5. Produce data on overall system performance under various conditions.
6. Demonstrate the total involvement of all branches of government and social institutions in the operation of the CJS to inhibit crime.
7. Serve as a test of the extent to which citizens, law-enforcement personnel, and elected officials can work together effectively in a continuing effort to make changes in social institutions as required by changing social conditions.

The combined effect of these several objectives is to find ways to systemize and improve the machinery of justice in all its aspects for the general benefit of organized society. The issue then becomes the determination of the essential requirements of a model criminal justice system. Initial work on the Laboratory County Program proposed these 10 requirements.

1. The system will deal with all impartially. The present CJS does not fulfill this requirement in law enforcement, the judicial process, or corrections.

2. Penalties for crimes should be modernized and made useful. The perpetuation of archaic laws that do not assist in rehabilitation is rejected. The penalties for crimes should be restated in operational language as much as possible, such as: to make restitution, to keep the peace, to protect the driving public against negligence until cured, and so on. Public acceptance of punishment as the primary goal of society in affixing penalties is changing rapidly, and other objectives for penalties should be defined and means of achieving them explored.

3. Immersion in the CJS should not destroy the accumulated assets of a citizen. By current practice, even an acquittal can bring financial ruin to anyone who challenges, in a legal setting, the inexorable patience and infinite resources of the government. Other potential citizen losses include reputation, respect, and the opportunity to prosper.

4,5. These two primary requirements of a model CJS deal with the importance of circumstance and intent. The circumstances of an offender should be taken into account—and that even when the crime is clear, penalties are agreed upon, and circumstances allowed for, penalties should not always be applied. Examples of circumstances and in-

tent are reasonable cause in apprehension, premeditation in trial, and health in corrections. These comments embody the principle that innocence and the hope of future virtue are more important than enforcing the letter of the law.

The sixth through tenth requirements become of concern only if the benefits to be gained appear worth the potential costs.

6. The system should deter citizens from committing criminal acts and should help to remove the forces impelling criminal activity. Included in this requirement is the elimination, where feasible, of root factors favoring crime. Motivations for a wide range of offenses should be curbed in any way possible.

7. The citizen should be safe from violence in his ordinary concerns. Either the CJS can, or cannot, provide protection within its budget. The balance between minimization of pressure on citizens by security forces and protection from criminal activities should be achieved, and alternative means to accomplish this should be explored.

8. The correction system should be rebuilt to maximize the potential value of convicted offenders who are returned to society. The expectation that convicted offenders will be worsened by their period of imprisonment must be reversed.

9. The system should allow no unnecessary delays and justice should not be subject to the whim of technicians. Many persons jailed for long periods of time have never been convicted of any crime. They are held

solely to ensure that they appear for trial. Other means must be found. The counter is also true; that is, technicalities provide escape routes for the guilty which were intended to protect the innocent.

10. The system should take some cognizance of the plight of the victim and some responsibility for redress. Crimes are often less hurtful to the state than to the victim, and in our tradition, criminal justice almost totally neglects the victim.

Ventura County plans to apply the systems approach to find answers to many questions regarding the CJS. A very detailed analysis and modeling program will be pursued. The specific activities that the laboratory county will undertake are:

1. A Ventura County crime and demographic profile.
2. A survey of other CJS.
3. A mathematical modeling feasibility analysis.
4. Goals will be defined.
5. Integration of the laboratory with the county.
6. The description of a model CJS.
7. A survey of subjective social factors.
8. The definition and categorization of system constraints.
9. Studies on the limitations upon the exercise of institutional and police discretion.
10. A study of the relationship between crime and criminality.
11. An analysis of CJS resource allocation.
12. The identification of further desirable social experiments.
13. The development of a model corrections facility.

Implicit in this description of the laboratory county has been the exposition of many issues associated with the CJS to which the systems approach can be applied.

References

(1) J. B. Jennings, Quantitative Models of Criminal Courts, Rept. P-4641, Rand Institute, New York, May 1971.

(2) A. Blumstein and R. Larson, Models of a Total Criminal Justice System, *Operations Res.*, 199–232, Mar.-Apr. 1969.

(3) J. B. Jennings, The Flow of Defendants Through the New York City Criminal Court in 1967, Rept. RM-6364-NYC, Rand Institute, New York, Sept. 1970.

(4) J. B. Jennings, Quantitative Models of Criminal Courts, Rept. P-4641, Rand Institute, New York, May 1971.

(5) J. Navarro and J. Taylor, Data Analyses and Simulation of the Court System in the District of Columbia for the Processing of Felony Defendants, *Task Force Report: Science and Technology,* President's Commission on Law Enforcement and Administration of Justice, Washington, D.C., 1967, 37–44, 199–215.

(6) *Final Report on the Development of a Criminal Court Calendar Scheduling Technique and Court Day Simulation,* Programming Methods, Inc., New York, Mar. 1971.

(7) F. Merrill and L. Schrage, Efficient Use of Jurors: A Field Study and Simulation Model of a Court System, *Wash. Univ. Law Quart.,* No. 2, 151–183 (1969).

(8) S. E. Kolodney and D. Daetz, Corrections Cost Projects, Rept. 458, Sylvania Electronic Systems, Mountain View, Calif., Jan. 1969.

(9) R. G. Chamberlain, Crime Prediction Modeling, Rept. 650-126, Jet Propulsion Laboratory. California Institute of Technology, Pasadena, Calif., Apr. 23, 1971.

(10) S. Angel, Discouraging Crime Through City Planning, Working Paper 75, Institute of Urban and Regional Development, University of California, Berkeley, Calif., Feb. 1968.

(11) J. Jacobs, *The Death and Life of Great American*

Cities, Random House, Inc. (Vintage Books), New York, 1961.

(*12*) P. Wormeli and S. E. Kolodney, Transaction-based Statistics for Criminal Justice Management Decision Making, presented at the 40th National Meeting, Operations Research Society of America, Anaheim, Calif., Oct. 27, 1971.

(*13*) P. W. Greenwood, Long Range Planning in the Criminal Justice System: What State Planning Agencies Can Do, Rept. P4379, Rand Corporation, Santa Monica, Calif., June 1970.

(*14*) R. Holbrook, M. King, and D. Webber, A Plan for the Development of a Laboratory of Criminal Justice, presented at the 40th National Meeting, Operations Research Society of America, Anaheim, Calif., Oct. 27, 1971.

CHAPTER 8

Urban Protective Services

This chapter on urban protective services will be divided into five major sections. The first section will look at general issues, especially those associated with the allocation problems of urban emergency units. The topics of the other major sections are police services, fire services, emergency medical services, and special warning services, such as earthquake and pollution.

Until recently urban emergency service systems have not benefited from a systematic analysis of their operational problems. The systems we refer to have the following properties (*1*).

1. Incidences whose time and place of occurrence cannot be predicted with certainty occur throughout the city and give rise to requests or calls for service.
2. In response to each call for service, one or more emergency service units, usually vehicles, are sent to the scene of the incident.
3. The speed with which the units arrive at the scene of the incident and deliver their service has some bearing on the actual or perceived quality of the service.

Examples of emergency units are police patrol cars, ambulances, fire engine and ladder trucks, tow trucks, bomb-disposal units, and emergency repair trucks for gas, electric, and water services.

The characteristics listed previously are shared by all the emergency services; however, the services differ in three significant ways (*2*). First, some units are located at fixed points while waiting for calls whereas others are mobile. Obvious examples are the mobile police cars and the fixed-location fire units. The impact of this difference influences both administrative and analytical purposes. Police patrol sectors must be designed very differently from the response areas of fire units. During periods of high demand the difference breaks down when it becomes necessary to dispatch units from one incident directly to the next without returning to any fixed location in between the incidences.

The second difference between emergency services is the urgency of the calls they receive. Coupled with this is their ability to detect the degree of the urgency in advance. Fake alarms are not urgent, but it is very difficult to identify in advance that an alarm is false even if meticulous records on previous false alarms are kept. If, at a fire, the fire chief calls for additional units, this call is clearly of high priority and very reliable. Since most criminal activity is reported by telephone, more detail is generally known. The dispatching policy of an emergency service is closely related to its ability to distinguish the priority of its calls. No units may be sent to a call, or a queuing procedure may be used if the priority is known to be low. If all calls are assumed to be urgent, and the demand is high, the options are to respond more slowly, to send fewer units, or to call in units from greater distances.

The third difference between emergency services is the nature of the time spent between servicing calls. For example, it is a belief represented by operational procedures that routine patrol by police cars acts as an inhibitor to criminal activity. Therefore, this important secondary function is a planned part of the police service. It may be desirable to place some calls to the police in a queue to preserve the crime-deterrent patrol.

The allocation or deployment policy of an emergency service system is the set of rules or procedures that determine the response patterns of units. Typically, the allocation policy is used to determine the following characteristics of the system:

1. The total number of specific units on duty at a given time.
2. The number of men assigned to each unit.
3. The location or patrol area of each unit.
4. The priority criteria to be applied to different types of calls and the circumstances under which calls are queued.
5. The number of units of each type dispatched to each reported incident.
6. The specific units to be dispatched.
7. The conditions under which the assigned locations of units are changed.
8. If units must be relocated: how many, which ones, and where.

The impact of the systems approach thus far has been to provide methods that can be used to select or improve the decisions made in relation to one or more of these components of the allocation policy. Other aspects of allocation policy will be examined first before looking at some of the systems allocation methods. Aspects of operations which influence allocation policy are the assignments given to particular individuals; the procedures followed at the scene of an incident; and the support functions of maintenance, supply, training, and administration.

Cities differ widely in their physical properties, and the choice of an allocation policy for emergency services must consider these properties. Such a property is the nature of the geographical area to be served. Geographic features will be part of the determination of the time it will take for an emergency unit to travel from one point in the city to another or to patrol a specified region. Other factors include the nature of the street network; the location of impediments to travel, such as rivers, railroads, and parks; the speed at which emergency vehicles can travel; and points of congestion. Emergency service agencies very rarely collect these data.

Allocation policy is also affected by the population density and land-use patterns. This kind of information gives the fire department an estimate of the threat to life and property by a fire in a particular location. The land-use information is indicative of the type of equipment that will be required to extinguish the fire. Police use the land-use data as an indicator of the number of doors and windows of commercial establishments which must be checked.

Statistical analysis of past calls will give the distribution of calls for service in a probabilistic form, and this is an important factor. Call arrival rates can vary by as much as a factor of 20 at different times of day and by a factor of 100 from one location to another. It is possible to make a general determination of these probabilistic properties which are common to all cities and to all emergency services. It has been found that the arrival times of calls can be described by a time-dependent Poisson process.

Some factors influence the performance that can be expected from any particular allocation policy. For instance, the number of units of each type necessary to handle each kind of incident and the length of time each has to work at the incident affects performance. For example, under one fire condition a single man with a portable fire extinguisher can put out a fire in 5 minutes. At the other extreme a fire may burn for several hours before being extinguished by the efforts of a ladder truck and an engine.

These factors raise the basic question of how to measure the performance of an allocation policy. Frequently the answer to this question is cast in the form of the specific emergency service being studied. However, a common general requirement is the optimal allocation policy for a predetermined total number of men, or the counter, that is, the overall number of men required to meet prespecified objectives. Contemporary costs have caused the fixed manpower restriction. An estimate of the cost of operating a two-man patrol car in New York City is $120,000 per year, and operating a single fire engine with its complement of men exceeds a cost of $500,000 per year. Adding emergency units with costs of such a magnitude quickly convinces administrators of the worth of an allocation policy that will get the most performance from their existing manpower.

The next five methods we will review are analytical techniques aimed at determining one aspect of the allocation policy, that is, the number of units to have on duty in each area.

1. The first method is based on geography and land use. The administrator of an emergency normally has little or no opportunity to relocate his service's fixed facilities. Each location was determined historically by the way in which the city grew. Therefore, part of the allocation policy is determined by the geography of existing facility locations. Insurance companies often reinforce the reliance on geographical factors by setting fire insurance standards based only on distance. For example, one standard requires homes to be no farther than 1 mile from an engine company and no farther than 1.25 miles from a ladder com-

pany. Any allocation policy set up this way must be regarded as inadequate. The time between receipt of a call for service and the arrival of units clearly depends on many other factors besides distance.

2. A second traditional approach utilizes workload or hazard formulas. These formulas were an attempt to introduce quantitative methods into a policy area which previously depended on judgment alone. An example formula, developed by Wilson (*3*), determined a police hazard score for each area based on several factors, including number of arrests, number of calls for service of particular types, number of accidents, number of doors and windows to be checked, number of street miles, number of licensed premises, and number of crimes. The hazard formula was defined in a linear form as

$$F_j = f_{1j} + f_{2j} \cdots + f_{ij},$$

where

f_{ij} = amount of factor j associated with area i
  ($i = 1, 2, 3, \ldots, j$ = number of factors),
F_j = city-wide amount of factor j.

To determine the hazard score for each area, weighting coefficients w_j were assigned to factor j to stress the importance, or lack of it, of the factor. For example, minor crimes would be weighted 1, more serious crimes 2, and very serious crimes 3. The hazard score for area i would be

$$H_i = w_1 \frac{f_{i1}}{F_1} + w_2 \frac{f_{i2}}{F_2} + \cdots + w_j \frac{f_{ij}}{F_j}.$$

Having computed the hazard score for each area, the total number of policemen would be distributed among the areas in proportion to the hazard scores. The flaws in this model are serious. The weights are subjective indicators of importance, allocations are frequently unsatisfactory, the linear model precludes description of the highly nonlinear and complex interactions possible, and deterministic definitions are used for variables which are likely to be probabilistic. In addition to these mathematical shortcomings, the formula does not relate meaningful measures of system effectiveness to operational policies. Nothing is said about the response time of units, the probability that a patrol car will intercept a crime in progress, or other operational factors.

3. Queuing theory has been applied to urban emergency systems where the consequences of placing a call for service in a waiting line or queue may be serious. The objective is to guarantee that every call will be able to obtain the immediate dispatch of a unit; however, when this is impossible, the objective of queuing analysis is to assure that the probability of an important call encountering a queue and waiting longer than some specified limit is very low.

The simplest generalization of the queuing model which has been applied to an ambulance service is of the following form. We assume that there are N vehicles located at a hospital and that each call requires the dispatch of one vehicle. If we further assume that all calls are identical in terms of their importance and service time, then from queuing theory the formula for the probability of the formation of a queue is known (*4*). The arrival of the calls is described by a Poisson process at an average rate of λ calls per hour. The service-time distribution is exponential with mean $1/\mu$ hours; that is, the probability of service lasting longer than t hours is $\exp(-\mu t)$. Under these conditions the probability of a call arriving when all N units are on service call approaches the value

$$p(N) = p_0 \frac{r^N}{N!},$$

where

$$P_0 = \cfrac{1}{\displaystyle\sum_{n=0}^{N-1} \frac{r_n}{n!} + \frac{r^N}{N!(1 - r/N)}}$$

and $r = \lambda/\mu$, which is less than N.

This mathematical model has provided a good description of the operation of ambulance services in many cities. If the administrator is given an estimate of the arrival rate of calls, he can choose the maximum value of the expected waiting time for service; then, using the formula, he can calculate the necessary number of ambulances to have on duty at various times. The City of St. Louis has used an identical model for the allocation of police patrol cars (5).

Queuing models have been extended to recognize that there are different kinds of calls. When the queuing model is applied to fire departments, the following characteristics must be built into the model.

a.   Different numbers of units of various kinds are required by different types of alarms.

b.   The fire units may arrive or depart singly or in groups.

c.   The type of fire determines the length of time the units are busy at the incident.

4. The queuing models just discussed are not adequate at relatively low alarm rates, where geographical requirements may dominate, or where services must perform important functions other than response to calls. In addition, the probability of encountering a queue does not have a clear relationship to the true performance of the system. Ultimately we would like to know the actual benefits gained by decreasing the delay of the arrival of an emergency unit. Possibilities are that stolen goods might be recovered, property damage averted, or lives saved. To use an analytical approach to the total response time, we define that time to be the period between the initiation of a telephone call or alarm and the arrival of the units at the scene of the incident. Therefore, the total response time includes travel time and any queuing, communications, and processing delays. Ideally we would like to have equations relating measures of response time to the number of units on duty, geographical features, arrival rates of alarms, and service times at incidents. Response-time models have been developed which show that the average travel time is inversely proportional to the square root of the number of available units, with the proportionality constant dependent on the travel speeds and locations of units. A very simple model shows that the average travel time for a single unit sent to an incident is approximately

$$T = \frac{0.63 \sqrt{A/N}}{v},$$

where

N = number of units available on random patrol,

A = area of region being patrolled,

v = response speed of the emergency unit.

Other results indicate that a reasonable approximation in cases where unavailable units are not common is to assume that the average travel time is inversely proportional to the square root of the average number of available units. With this information the following steps will determine how many units to have on duty in a specific region:

a. Using data, compute the constant of proportionality between average travel time and the inverse of the square root of the average number of units on duty.

b. Estimate the arrival rate of calls.

c. Using queuing models, compute the probability of a queue for each number of units expected to be on duty.

d. Using results a and c, estimate the average travel time.

e. Select the number of units on duty to be large enough that a threshold waiting time is not exceeded, and that the average travel time does not exceed another specified threshold.

This response-time method gives a good approximation to the desired allocation for each region.

5. Beyond the analytical techniques just discussed are others designed to take into account two or more performance criteria. A dynamic programming model has been developed by Larson for allocation of patrol cars to precincts (6). The algorithm utilizes a queuing model, a travel-time model, and a model of the frequency with which cars pass by an arbitrary point while on preventive patrol. The strategy of the allocation is to state the policy objectives in terms of constraints. For example, it may be decided that the average travel time should not exceed 4 minutes. If a set of constraints is unobtainable, the algorithm specifies the additional resources necessary to meet the objectives. An allocation policy developed with the algorithm is much more operationally oriented than the hazard formulas discussed previously.

Another interesting queuing model was developed by Carter and Ignall (7). Its function was to determine the extent to which an additional fire unit provides relief to overworked units in its area.

The application of any of the five sets of allocation models which we have discussed requires a great deal of effort from any staff. A number of quantitative decisions must be made which require very specific information from the concerned agency. The appropriate determination of what constitutes an "excessively long" delay before the arrival of a unit, how much preventive patrols will be considered adequate, or what level of workload is "too great" must be done. Beyond the design of the allocation policy by the methods presented lies the difficulty in assigning individuals to shifts or tours of duty who best fit the assignment. This issue is important and little work has been done to resolve it. Other important related problem areas are the design of response areas, locating units and facilities, and design of patrol services. The references cover these research areas in some detail.

A complete allocation policy is very complex

and the evaluation of proposed change in the policy is important. For example, an analysis may suggest that more emergency units should be added for certain times of day, new locations for some units, different response or patrol areas, or modification of the procedures for relocating units. Any administrator will want to have a realistic comparison of the benefits that can be expected from any change in the allocation procedures. Large-scale simulation models have been used for this purpose. Such simulation studies have investigated the reduction of travel times which could be achieved by spatially repositioning ambulances, have analyzed the operations of the dispatch centers of fire departments, have compared a wide range of combinations of fire department allocation policies, have evaluated the potential benefits of utilizing a car location system, and have studied the value of an automated dispatch system for the police department.

One of the most interesting results of these studies has been to learn that simple and inexpensive administrative innovations can often make a contribution to system performance which is equivalent to that of much more expensive hardware or increases in manpower. For example, simulation has shown that the absence of an explicit priority structure for calls to police departments produces unnecessary delays for urgent and moderately important calls. The police department's position is that they want to provide rapid service to all citizens; therefore, there is resistance to the implementation of a priority structure, even though their idealized position does not work well.

Simulation studies have also been used to evaluate the use of automatic vehicle-locator systems in police departments. These systems are already well developed and they provide a central dispatcher with estimates of the positions of all service units, including their current speed, direction, and type of activity. Analysis showed that the combination of nonoverlapping patrol sectors and an automatic vehicle-locator system caused an average travel-time reduction in the order of 10 to 20 percent.

Police Services

Decision Making in Police Patrol

Most problems associated with urban police forces are related to a growing demand for services without the appropriate increase in allocation of resources. There is a need for effective aids in decision making regarding many facets of police services. Recent research pointed out decision making areas in police patrol as (8)

1. Proper patrol-force strength.
2. Equitable and effective distribution of patrol services by police district and tour of duty.
3. Effective operational policies and tactics for police patrol.

The importance of police patrol is substantiated in these figures. Annually we spend over $5 billion on our CJS, with over half going to the police agencies. Well over $1 billion is allocated to police patrol—the heart of the police law enforcement.

Systematic evaluation of police patrol operation is difficult, because although police departments collect extensive data on each crime, they often fail to collect and use data that are relevant for the management and evaluation of patrol operations. When evaluation criteria are employed, they vary widely in different jurisdictions. If we were to idealize evaluation criteria we would choose true crime rate, true victimization rate, total social and economic impacts of crime, number of crimes prevented and deterred, and criminal and public attitudes in relation to alternative police programs. Obviously, however, these criteria cannot be easily or accurately measured at this time. Proxy measures must be used; the following set of police-patrol-evaluation criteria appear to have wide applicability:

1. The proportion of reported crimes for which at least one suspect is arrested (reported by crime category).
2. The proportion of reported crimes for which at least one suspect is formally charged in the judiciary branch of the CJS (reported by crime category).
3. The reported number of crimes per thousand population, by crime category and citizen group.
4. Reported crime by crime category.
5. Percentage of citizens satisfied with various aspects of patrol service.
6. Measures of the average time and times exceeding a threshold of the response times of police service.
7. The number of times a patrol car passes a randomly selected address per unit time.
8. Hours of preventive patrol per crime that may be significantly deterred or prevented by police patrol activity.
9. Resources expended, that is, total patrol man-hours and total car-hours, plus a break-

down of each total into percentage allocated to each patrol function.

In order to implement these criteria more data are required than are ever available. In the study we have cited, information was gathered from six major police jurisdictions, which included departments ranging in size from medium to large, employing many patrol allocation techniques, and the sample included older and densely populated as well as newer and more sparsely populated cities. In addition to the data normally available, the following detailed information was collected: crime, calls for service and deployment, trends overtime of demands for police service and patrol manning, patrol organization and operational policies, the internal police planning organizations, the uses of data and computers in police departments, and criteria and methods of allocating patrol resources. Some of the interesting results of this data collection were the wide variations of significant dimensions of police departments, such as:

1. Crimes per capita vary from 0.03 to 0.14.
2. Police budget per capita varies from $13 to $85.
3. Uniformed patrol strength per 1000 residents varies from 1.0 to 2.5.
4. Patrol cars on the street per shift per square mile vary from 0.045 to 2.7.
5. Calls for service per patrol car per shift vary between 2.5 and 11.5.
6. Crimes per uniformed patrolman vary between 26 and 70.

One of the observations that can be made from these data is that increases in police strength have not kept pace with increase in reported crime per capita, although they have outpaced population change. All six departments use the computer for one or more of the following: arrest, crime, incident, narcotics, clearance, personnel, and traffic reporting.

Most departments do not collect the management data that would be useful in providing relevant inputs into the major patrol allocation and tactical decisions we posed earlier. Command discretion and judgment dominate all decisions involving patrol force; however, a variety of quantitative allocation aids are in use. Many versions of hazard formulas are used, and we have already discussed their limitations.

One of the most advanced operational methods of patrol force deployment has been developed and applied in St. Louis. A prediction of the demand for police service is made by hour and geographic area. Then a mathematical technique is used to estimate the number of patrol cars required to answer without delay 85 percent of all the incoming calls. Fifteen percent of the patrol force is assigned to preventive patrol.

The study of police patrol decision making by Kakalik and Wildhorn (8) offered the following suggestions:

1. Employ multiple criteria in decision making.
2. Utilize modern quantitative allocation methods of the type we described earlier.
3. Collect certain management-oriented data.
4. Hire civilian systems analysts and planners and give them access to top police management.
5. Perform research and experimentation so that the relationship between police resource

inputs and police effectiveness is better understood.

Beat-Design Problem

We mentioned earlier that the design of patrol facilities was a significant factor in the allocation of police resources. In this section we will see how the staff of the St. Louis Police Department partitioned their police districts into patrol beats which provide rapid response to calls for service, are efficiently patrolled, and have equal shares of the workload (9).

The design of police patrol beats is similar to forming political voting districts. Political voting districts are formed by grouping census tracts or other population units to form compact, contiguous, voting districts which are roughly equal in population. In the same way, police patrol beats are groups of reporting areas which form compact, contiguous areas that are roughly equal in workload and in crime load. The objectives of the design are to minimize the maximum travel time between any two points within a beat and to control crime. The specification of a crime load in a reporting area is difficult because the times of occurrence of many crimes is not precisely known, some crimes are more serious than others, and many crimes are never reported. For the first problem the time of reporting is substituted, and for the second, an index of seriousness may be used or only the most serious crimes are counted.

The ease with which a beat can be patrolled depends on how the streets within it run, how many streets are thoroughfares, what obstacles to the flow of traffic such as railroad tracks or waterways are present, and whether there are known trouble spots in the beat.

In order to accumulate police data, St. Louis has been partitioned into 490 reporting units known locally as Pauly blocks. Each Pauly block contains 9 to 12 city blocks. Data on police service calls are kept and used to produce a forecast of average weekly demand by Pauly block. Weighting factors are used to break down the weekly demand to daily patrol forecasts.

Computer programs have been developed for political districting which use the population moment of inertia of voting districts about their population centers as the measure of compactness. District centers are formed in such a way that the population per district falls within certain bounds and the sum of each district's moment of inertia about its center is minimized. The contiguity of districts is then checked. Therefore, the algorithm of the program generates all feasible districts which are contiguous, compact, and have limited population deviation, then selects the best set of districts which covers all the census tracts exactly once.

The modification of this algorithm for the design of patrol beats is fairly direct. Pauly blocks are used instead of population units, the forecast workloads in the form of service times replace the population statistics, and the compactness and contiguity concepts are the same. The equal-crime-load requirement is brought in by adding a feasibility constraint and providing a data table giving the crime load for each reporting area. The last constraint added specifies the acceptable range of deviation of a beat's total area from the average. The shape compactness is defined in terms of the d^2A,

where

d = distance compactness, which is measured by the distance between the centers of the two population units in the district which are farthest apart,

A = area of given district.

All other constraints for the program may be written in the form

$$| V(n) - V_a | \leqslant fV_a,$$

where

$V(n)$ = one of the attribute values for the nth beat such as workload, crime load, or area

V_a = average attribute value taken over the beats,

f = allowable fractional deviation from the average.

Using these constraints, the algorithm will formulate all optimal beat plans, where optimal means minimized maximum deviation of any beat's workload from average and where each beat meets the contiguity, compactness, crime load, and area constraints. Data taken from a St. Louis police beat were used to validate the computerized beat-design algorithm.

Incidence-Seriousness Index

Another useful quantification device developed by the staff of the St. Louis Police Department is a means of measuring the seriousness of crime by the use of an incidence seriousness index (*10*). An evaluation of crime would be of great value in the allocation of police department resources. Even more important is the picture of crime which the index allows. That is, viewing the fine structure of criminal activity—the physical injury, property loss, and intimidation—permits some remarkable conclusions regarding the nature of criminal activity.

The seriousness quantification technique is adapted from the quantitative index of delinquency developed by Wolfgang and Sellin (*11*). They established scores for each component of crime. As an example, their evaluation of the injury component of victims who have been assaulted is given next.

Injury Component	Score
Minor injury	1
Treated and discharged	4
Hospitalized	7
Killed	26

A measure of seriousness is given by the sum of the seriousness scores for the relevent events, multiplied by the number of times each has occurred. The Wolfgang–Sellin index has a number of advantages and disadvantages reported in the literature; however, it proved very useful in this research. A system was designed to produce distributions of crime seriousness by watch, day of week, and police district. The distributions are displayed in a report that presents, for a specified class of offenses, seriousness information according to the categories of injury, intimidation, and property loss. Some of the results of the use of this system are given next. It must be pointed out that no other data-reporting system permits these kinds of conclusions.

1. Two-thirds of the harm from crime may be attributed to property loss, and one-sixth each to physical injury and intimidation.
2. The average seriousness of a crime against the person is more than four times as great as the average seriousness of a crime against property.
3. Crimes that ˙occur in places not readily viewed by police on patrol are substantially more serious than those taking place on the streets.
4. Crimes against the person accounted for 12.5 percent of the incidents but 37.5 percent of the seriousness.
5. If the seriousness of crime is assumed to be an important factor in allocating police manpower, then some commonly used methods of allocation may actually be misallocating resources.

Using the data from a total of 9827 offenses over a 2-month period, the average seriousness was found to be 3.00 units, consisting of the following components:

Injury	0.49 unit	15.9%
Intimidation	0.53 unit	17.7%
Property loss	1.99 units	66.4%
	3.00 units	100.0%

Many of the study results were directed to manpower-allocation problems. The conclusions include the following:

1. Deploy preventive patrols in proportion to the distribution of the seriousness of suppressible crime.
2. Deploy small, special-purpose squads deal-

ing with a specific type of crime according to the distribution of the seriousness of the crime.

3. Revise the workload formula to allocate patrol manpower to districts so that the weighting factors reflect the seriousness of the related incidents.

4. Design patrol beats so that, in addition to balancing the expected workload in each, the total seriousness of crime in each is balanced.

5. Modify the queuing models so that the average response delay is reduced when the average seriousness of the crime reported is above the city-wide average.

Another hope is that crime seriousness data may prove useful in measuring the effectiveness of police operations. Seriousness-based measures of effectiveness have been suggested for three types of activities: intervention in crimes in progress, apprehension of offenders, and criminal investigation.

Application of PPBS to a Police Department

In Chaper 2 we reviewed the methodology planning–programming–budgeting system (PPBS). In this section we will see how PPBS will help police planners examine alternative courses of action in the light of the costs and benefits of such actions and police department objectives and goals (*12*).

Recall that the principles of PPBS require the identification and definition of an organization's objectives followed by grouping the organization's activities into programs that can be related to each objective. The objectives of a police department are not adequately stated as to protect lives and property. Instead they must be identified more particularly as apprehending criminals or directing the flow of traffic on city streets. A program that has an activity intended to accomplish an objective always has an end product that can be taken as a measure of the program's effectiveness. Some end products are difficult to measure quantitatively or even identify unambiguously. Benefit is an indicator of the utility to be derived from each program. The program budget presents resources and costs categorized according to the program or end product to which they apply. A conventional budget gives the budgetary needs by type of resource, such as cars, personnel, desks, gasoline, or buildings, or by functional category, such as precinct commands, communications bureau, or motorcycle squads. The point of PPBS is to focus resources toward competing programs and on the effectiveness of resource use within programs.

In the application of PPBS to the New York City Police Department, the following programs and their goals were identified.

1. Crime prevention and control—to prevent crimes, to intervene in criminal activity, to detect and arrest offenders.

2. Investigation and apprehension—to identify, discover, and apprehend suspected and known criminals, to control criminal activity, to provide internal security and other functions.

3. Traffic control—to enforce traffic regulations and to control vehicular traffic.

4. Emergency services—to provide immediate

emergency and rescue assistance at scenes of serious incidences, and to help prevent such incidences through patrol.

5. Support—to provide a variety of administrative services, including base facilities, license issuance, and recruit training.

An indication of the importance of the five programs is the percentage of manpower commitment and total cost per program:

Program	% Manpower	% Total Cost
1	67	68
2	11	14
3	13	9
4	2	2
5	7	7

The procedure now is to examine each program package to find elements directly related to the mission that could be differentiated and substituted for one another. Program 2, for example, contains six subprograms: administration, special squads, borough and precinct commands, technical services, investigation and indictment, and investigation of municipal affairs. Note that no subprogram necessarily supports the others, although each supports the Program 2 mission.

We will continue with Program 2 as a running example and identify the existing organizational units.

Program Structure Elements	Existing Organizational Units
Administration	
Administration	Chief of detectives' office and staff
Special squads	
Special squads	Chief of detectives' patrol squads
Homicide	Borough homicide squads
Burglary	Burglary and larceny bureau and squads
Youth	Youth aid bureau units
Narcotics	Narcotics bureau squads
Riverfront	Riverfront squad
Missing persons	Missing persons bureau personnel
Central Investigation bureau	Central investigation bureau personnel
Bureau of special services	Bureau of special services personnel
Borough and precinct commands	
Administration	Detective borough and district commanders and staff
Operations	Precinct detective squads and borough youth squads
Technical services	
Laboratory	Scientific research unit
Ballistics section	Ballistics section
Bomb squad	Bomb squad
Photographic section	Photographic and photostat unit
Investigation and indictment	
Bronx	Bronx District Attorney squad
Kings	Kings District Attorney squad
New York	New York District Attorney squad
Queens	Queens District Attorney squad
Richmond	Richmond District Attorney squad
Investigation of municipal affairs	Department of investigation squad

The next step in PPBS is to develop the cost structure, which uses cost elements to specify costs. In this way the use and allocation of resources is more visible. The following cost structure was developed for the cited study:

Initial Investment (Nonrecurring Costs)	Annual Operating (Recurring Costs)
1. Facilities	1. Personnel
2. Transportation equipment	2. Facilities operation and and maintenance
3. Command communication equipment	3. Transportation equipment
4. Other equipment	4. Other equipment
	5. Administration
	6. Debt services

Each of these 10 items of the cost structure can now be identified with line items in the current conventional budgeting system. No line items in the budget are divided between two or more cost elements.

Methods must be established now by which cost can be estimated in program terms. The level of detail is an important consideration. For example, data relating to the Personnel cost element were easy to obtain, and because this element accounts for over 91 percent of the total cost of the police department, the cost element was estimated all the way down to the lowest level of program detail. On the other hand, Facilities and Debt Service were estimated only at the program level, because data were not available. Program 2, Apprehension, is shown in the next table, where an × indicates the level of costing detail.

	Program Level	Subprogram Level	Program Element Level
Facilities			
Transportation equipment	×	×	
Command and communications equipment			
Other equipment	×	×	×
Personnel	×	×	×
Facilities O. and M.	×	×	×
Transportation O. and M.	×	×	
Other Equip. O. and M.	×	×	×
Administration	×	×	×
Debt Service			

This table must be filled in by the cost estimates for each program category. Cost estimates for the items of the cost structure for each program are determined by a variety of means. Some cost items require a separate extensive methodology as in the case of Personnel. As an example of the cost estimates, the personnel item is given next for Program 2, Apprehension, in thousands of dollars.

Administration	255.0
Special squads	18,014.6
Borough and precincts	30,845.0
Technical services	5,580.7
Investigation and indictment	3,246.1
Investigation of municipal affairs	302.9
Apprehension total	58,244.3

The complete program budget for the police department will be the collection of all cost items for each program.

Remaining in the PPBS approach is to reconcile the program and conventional budget item by item; in this way, the items that depart from the PPBS philosophy are on display and can be examined carefully. In reference (*12*) the proposed conventional budget for the New York City Police Department was $551.6 million and the PPBS budget was estimated to be $488.3 million or $63.3 million less than the conventional budget estimate. Comparing the two approaches for the personnel item, we have

Conventional budget	$496,715,400
Program budget	431,045,600
Difference	65,669,800

Each point where the two budgets depart is studied until the difference is accounted for totally. For example, one item in the conventional budget is a cost of $39,240,000, which comprises payments from the old pension system. This is a nonrecurring cost and should not be included when estimating future personnel costs of the department.

The final ingredient is to develop a computerized cost model to assist in examining alternative decisions with sustained cost implications. This was done in the language of both the program structure and the cost structure of the program-budget format in order to make cost projections for one program or subprogram at a time. The output of the model presents static costs, which can be used throughout a study to analyze the effect on total system cost of possible changes in equipment design and in system operation.

Fire Services

Urban Fire Protection (*13*)

The mission of fire departments everywhere traditionally has been to prevent fires from occurring and to respond to those that do occur and put them out. In spite of the stress on the preventive role, fire departments are organized primarily to serve in crisis. Since fire is inherently a physical and chemical phenomenon, there is a substantial technological orientation from prevention to extinguishment. Fire service functions include inspection or advice on building plans; inspection of industrial, commercial, public, and residential buildings; the actual extinguishing of fires and putting the property back in order; and a public education program. To properly discharge these functions requires the cooperation and coordination of a variety of persons and agencies: architects; building contractors; those who formulate and administer codes; fire insurance companies, whose ratings influence the use of detectors, sprinklers, and the clearing of brush; telephone companies; alarm services; and equipment manufacturers and suppliers. This outside network is an auxiliary fire-protection system, in a sense, and when they perform poorly, they impair the fire service's ability to carry out its mission.

Fire departments are essentially line-operating agencies and, as a part of the bureaucracy, are often poorly equipped in outlook, skills, and organization to undertake novel or significant change. Conditions are such, both organizationally and politically, that the rewards for success are too small and the price of failure disproportionately high. Little support is received by fire departments from industry, universities, or the federal government. Instead, information and new ideas come largely from a few dedicated special-interest groups and professional associations, which are supported by limited financial resources and little research. For these reasons, and in spite of new types of equipment, basic fire department practices seem to have changed little in the past century. Mobile radio equipment and power tools are beginning to be used more extensively, but new materials, new protective clothing, new fire detectors, new extinguishing agents, or materials to improve the effectiveness of water have been introduced very slowly. This is due, in part, to the small and distributed fire service market and a fragmented and reluctant supply industry.

As with all urban protective services, the demand for service has grown much more rapidly than the resources. In New York City between 1956 and 1969 fire-alarm rates more than tripled. The rate for every type of incident has been increasing exponentially. Curiously, false alarms, rubbish fires, nonfire emergencies, fires in vacant or abondoned buildings, and deliberate fires now outnumber the accidental structural fires that used to be the fire department's main effort.

Four areas in which recent research has been accomplished are communications, deployment, management information and control systems, and new technology. The communications effort has emphasized the dispatching centers which form the link between alarms from the public and the department's fire-fighting response. The two aspects of dispatching are receiving, interpreting, and identifying alarms; and allocating and dispatching fire-fighting units to respond. A study of dispatching that included developing a model revealed a bottleneck at the point where response decisions are made and carried out. Simulation studies showed that increasing the number of dispatchers or the use of a computer had little effect. Instead operations were divided at the critical decision, and this effectively doubled the system's ability to handle peak loads. The Brooklyn installation that used this idea did so at a cost of less than $1000.

Another research effort has been directed to the allocation of fire-fighting units, that is, how many units to have and how to man and deploy them. In the past most fire departments have given a uniform "standard response" of men and equipment to alarms in most areas at all times. This is in spite of the fact that the frequency of alarms is much higher during certain periods, and fire hazards and the likelihood that an initially indeterminant alarm will turn out to be a serious fire vary greatly with area and time of day. In addition, the number of men and equipment deployed varies from city to city.

Many options were available, but analysis showed the following options to be the most useful. Because alarm rates are twice as high during the 3:00 P.M. to 1:00 A.M. period as any other time, and the types of incidences statistically predictable, an adaptive response was developed. The number of men and equipment dispatched were varied depending on the likelihood of given types of alarms and hazards at various locations and times of day. Another model pointed out the fallacy of a traditional dispatching rule which dictated always sending the units closest to an incident. Analysis showed that when nearby units have widely different workloads, other dispatching assignments dominate this traditional rule. As a result, response areas have been redesigned for battalion chiefs. The combination of these two deployment innovations would have cost the New York Fire Department an additional $5 to $15 million per year if supplied traditionally.

Two research impacts on the technological aspects of fire fighting have been early-warning fire-detection systems and "slippery" water. The detection systems use ionization detectors to detect thermal or smoke sources in buildings to save lives and sharply limit the number of serious fires. Slippery water refers to the effect of adding a special chemical to water which allows an increase of 70 percent or more of the flow through a hose at a given pressure. Smaller, lighter, more manuverable hoses can be used to increase the effectiveness of fire fighters.

Predicting the Demand for Fire Service

We would like to use systems analysis to provide a method for short-term prediction of the incidence rates for various types of fire alarms as a function of location, time, method of reporting, and weather conditions (*14*). There are records for 1½ million alarms and the effort will be to extrapolate past data, not develop causal statements. The model begins by assuming that given the type of alarm and method of reporting, the occurrence of that type of alarm is an independent Poisson point process whose parameters vary with time, geography, and weather. The predicted number of *a* type of alarms is taken to be

$$P_a(t,w) = P_a(t) + f\big[O_w(t-s) - P_w(t-s)\big],$$
$$s = 0, 1, 2, ..., k,$$

where

$P_a(t)$ = predicted fire alarms of type a at time T given normal weather,
$P_w(t)$ = predicted weather at time t,
$O_w(t)$ = observed weather at time t,
f = function of residual weather, which may also depend on values prior to t.

To compare $O_a(t)$, the actual number of alarms, to $P_a(t)$, the predicted number of alarms, by the method of least squares, we look at

$$O_a(t) - P_a(t) = \sum_{s=0}^{k} \gamma_s \big[O_w(t-s) - P_w(t-s)\big]$$
$$+ \text{ error}$$

for fitting γ_s.

To account for geography, let $X(t, r, j, i)$ be the number of type j alarms received in time unit t by reporting method r, at location i, and let

$$X(t, r, j) = \sum_{i=1}^{I} X(t, r, j, i).$$

If both $X(t, r, j)$ and $X(t, r, j, i)$ are assumed to be Poisson random variables with parameters $\lambda(t, r, j)$ and $\lambda(t, r, j, i)$, respectively, then let

$$p(t, r, j, i) = \frac{\lambda(t, r, j, i)}{\lambda(t, r, j)}.$$

From this it may be shown that the random vector

$\left[X(t, \quad r, \quad j, \quad 1), \quad \ldots, \quad X(t, \quad r, \quad j, \quad I)\right]$ has a multinomial distribution with parameters h and $p(t, \quad r, \quad j) = \left[p(t, \quad r, \quad j, \quad 1), \quad \ldots, \quad p(t, \quad r, \quad r, \quad I)\right]$. Determining the geographical effect on time patterns amounts to testing whether $p(t, \quad r, \quad j)$ depends on t and can be done via a chi-square test on each contingency table, given r and j.

The model we have presented is in an early stage of development and results thus far are mixed; however, it is an interesting example of a probabilistic prediction model.

Fire Service Location

In earlier chapters we have used location theory in a variety of ways. In this section the location of fire services is examined. Earlier in this chapter we literally dismissed the location of fire stations as out of our control, because they were added at heavy demand points as the cities grew. Now as parts of the city decay and are rebuilt, the opportunity is present for an optimal relocation of many fire stations (*15*).

An interesting study was done by Hogg in 1968 when plans were made to virtually flatten and rebuild central Glasgow by the 1980s (*16*). This was a rare opportunity to optimally locate the fire stations. Hogg chose the objective function of minimizing the total journey time of all the engines traveling to fires, including both first-responding engines and any reinforcements. The response time to a fire depends on where the fire occurs, where the nearest station is located, and the nature of the intervening roadway. Her procedure was as follows:

1. Designate an initial set of sites, evenly scattered over Glasgow, and satisfying any political constraints.
2. Depict the projected fire-incidence pattern in 1980 using regression analysis.
3. Divide the city into subareas composed of several 1-square-kilometer cells of the map grid.
4. Compute the travel time between the subareas and all possible station locations.
5. From all possible sites, the best were selected for the total number of engines (*41*).
6. Then the one site which increased the total response time by the least amount was eliminated, leaving 40 engines. The procedure was repeated and the previously rejected site (number 40) was compared with each of the retained sites. If an exchange could be made that would minimize the total journey time, one was made. The same analysis was made for station number 39 and the procedure continued until no sites were rejected.

In other studies where the fire station locations were fixed, optimum redistricting was performed by systems analysis methods. Some of the many location models which have been developed are: minimize response time, minimize response distance, an availability model in which the nearest unit is not assumed to be always available, a model that relaxes the assumption that only one unit is required on all alarms, a model which assumes that there is a maximum allowable travel time associated with each incident, and a model that attempts to balance the workloads while minimizing the total travel time.

Simulation of Fire Service Operations

We have discussed several attempts of fire service operations, and, as before, the simulation methodology offers a means of examining the fire service as a total coordinated operation. Before we do this, however, it is appropriate to list questions that a simulation might help us answer (*17*). First, the long-range questions:

1. What are the real costs to society of both the direct and indirect (loss of employment, increased insurance rates, and so on) effects?
2. Where can the greatest cost reductions be made for the total cost of the fire burden?
3. What are the actual costs of fire prevention and protection to society, when we consider both the direct and indirect (loss of tax revenue, high-pressure water supply, and so on) costs?
4. What is the optimum level of fire service to minimize the actual cost of the total fire burden, or to minimize the measurable total cost of the fire burden?
5. What is the relationship between the level of performance of the fire service and the loss of property and life?
6. How are the future requirements of fire service determined to assure a given service level when we expect changes in population, housing, and industry?
7. How do alternative building and safety codes and public education programs affect fire prevention?
8. How will changes in the number of fire boxes, water supply, and equipment maintenance policies affect fire-fighting services?

9. What are the best procedures for extinguishing fires?

And among the short-range questions to which simulation studies can contribute are:

1. How should fire-fighting equipment and personnel be distributed to minimize loss of life and property?
2. What are the optimum number and location of fire stations?
3. What are the best geographical areas to be assigned to each fire station?
4. How should fire equipment and personnel be assigned effectively to different types of fires?
5. What is the effect of over- or undercommiting resources on the first response to an alarm?
6. What is the most desirable allocation of men per shift to assure maintenance of the desired service levels?
7. What are the best strategies for using the fire service force for situations such as widespread arson?

We have considered solutions to several of these questions, and practitioners of systems methodology are dealing currently with many more. A simulation developed by Carter and Ignall was designed to compare different policies for locating, relocating, and dispatching fire-fighting units (*18*). The simulation allowed comparing proposed solutions to workload and response problems. The proposed solutions involved creating new units, some of them to work only in peak-alarm-rate hours. Reference (*13*) cited these results. This simulation emphasized the use of internal measures of performance.

In another useful study, queuing, fire station location, and simulation models were applied to sample data extracted from the historical records of the Alexandria, Virginia, Fire Department (*19*). A fire alarm was referred to as a case; as simulated time progresses, cases enter the system. The assignment portion of the model acts as the dispatcher by determining which particular resources will service a case. These resources travel to a case, remain there for a designated period, then leave the scene and return to their stations, thereby terminating the case. Features of the simulation model that were not necessary, but may be required to simulate other fire departments, included queuing of the cases by the dispatcher, delayed response time when resources are occupied, interruption of a low-priority service when an urgent case occurs, and temporarily transferring resources into other stations when necessary.

Another simulation that has fascinating prospects is the simulation, using a computer, of fires in buildings (*20*). The fire simulation requires that data about the building be assembled, the data arranged and stored, the fire described, and the interaction between the building and the fire analyzed. The most difficult and, at this point, incomplete part of the analysis is the representation of air movements in the building. The problem of computing air flows is that the equations describing the flow are in the form of elliptic partial differential equations, or a set of simultaneous algebraic equations for which solutions are long and difficult to obtain. Enough accuracy is possible in the model now to design sprinkler systems for some buildings. An exciting prospect would be a fire truck with input and output computer terminals so that an existing fire could be analyzed at the scene, and optimal strategies for fire suppression developed and implemented.

Emergency Medical-Care Services

State of the System

The contrast between medicine's lifesaving techniques and the system organized to deliver them is substantial in many ways. Nowhere is the difference more sharp than in emergency medical systems in most cities and nearly all rural areas. These systems do not often enough bring the patient at the right time to the right place; nor is there assurance that the patient will get the proper treatment by the proper professional. The many fact-finding studies and surveys that have been conducted by the Public Health Service and other groups have resulted in these observations (*21*):

1. Larger cities usually have governmentally provided emergency ambulance services primarily for street and police-call cases. Forms of emergency services include: hospital-based, police department, volunteers, fire department, and city and county health departments.
2. In smaller cities, with one or two hospitals, ambulance service is usually provided by funeral directors or volunteer groups.
3. Many cities do not have adequately equipped emergency vehicles.
4. Ambulance personnel often do not have suitable training.
5. Many hospital emergency departments, especially those in small communities, are not adequately prepared for handling emergencies.

6. Few communities have regulations relating to the operation of ambulance services.

7. Seldom is there municipal recognition for the need of an emergency medical services committee to supervise the total activity.

8. Where properly organized, volunteer rescue squads have a high sense of community responsibility and perform a good job.

9. Economic necessity is causing more funeral directors to withdraw from this type of medical service.

10. In some cases, proprietary ambulance companies provide excellent service.

The physician takes the viewpoint that an emergency need exists in a patient's mind until he is convinced by a health professional that his life or well-being or that of a loved one is not threatened (22). Therefore, professional decisions are required to determine the time, the type, the amount, and the place of emergency treatment. The factors and functions of the emergency medical care system are depicted in Figure 8-1. The elements vital to this system are planning, manpower, research, standards, and education.

The potential problems with the system in Figure 8-1 are delay in treatment, wrong treatment, a combination of both, inadequate facilities, and or a lack of knowledge. The problem to which systems analysis has been directed is delay in treatment. When there is a delay in treatment, it may be due to delays in discovery, in transportation, or in arrival at a proper facility.

In Belgium, because of increasing traffic accidents and difficulties in discovery of the incidences, emergency telephones were placed at 1-mile intervals along the road between Brussels and Ostend. Each of the telephones was tied into a communications network designed to send an ambulance promptly to the scene of an accident. Delay in discovery of accidents in an urban setting occurs primarily in fringe areas, obscure locations, and at unusual hours.

A delay in transportation relates to the now-familiar minimum response time. The difficulty may be communications, dispatching, or travel time. Delay in reaching the proper facility is in-

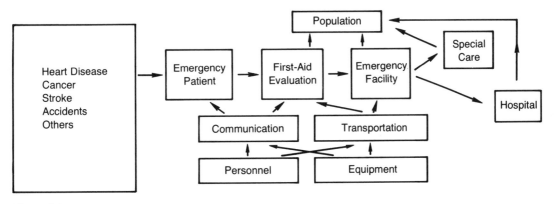

Figure 8-1

volved generally with inadequate planning of facilities.

Simulation of Emergency Ambulance Service

The simulation study that forms the basis of this section was of the emergency ambulance service in New York City (*23*). The study is limited to a consideration of the dispatching of ambulances in response to calls. When the study was initiated, all ambulances were located at hospitals, and the issue to be decided was the value of establishing satellite garages for ambulances in the vicinity of areas from which large numbers of emergency calls originated.

All calls for emergency service are routed to communication centers maintained by the Police Department of New York City. The calls for service originate with patrolmen or by the police using special telephone numbers. A police officer determines the district in which the emergency is located and contacts a dispatcher at the appropriate hospital by regular telephone to request that the ambulance be sent to the incident. If no ambulance is available, a neighboring district is contacted.

The simulation model was based on the emergency experience of a particular hospital, the Kings County Hospital. The model was constructed so that emergency calls could be generated randomly with a frequency and geographical distribution corresponding to observed data. The distribution of arrival calls is taken to be Poisson, and the model is formed so that any number of ambulances can be stationed at the hospital or at any number of satellite garages. The cycle of events for the movement of ambulances is illustrated in Figure 8-2.

Conditions were simulated for the heaviest tour of duty. A daily average for the hospital was one call every 500 seconds. Information about travel time was derived from observations of the movements of two ambulances on call for a period of 2 weeks. For a large percentage of the calls, travel time was proportional to distance. The data were fitted by regression analysis with two straight lines; one covering calls up to a distance of 3.4

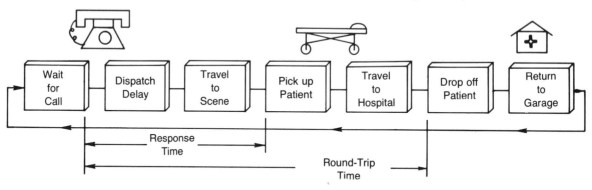

Figure 8-2

miles at a speed of 15.4 miles per hour. The other for distances beyond 3.5 miles for a speed of 37.4 miles per hour. On the basis of the straight-line representation the average travel time t, in seconds, was calculated from distance d, in 1000-foot units, by the following two equations:

$$t = 44.3d, \qquad\qquad d < 18.48,$$
$$t = 18.2(d - 18.48), \qquad d > 18.48.$$

When an emergency call came in, the computer program searched through all the ambulances at the hospital, at satellite garages, or on the street returning from an incident with no patient. The nearest ambulance was assigned to the call and the travel time computed. The arithmetic average of the pick-up time was found to be 7.62 minutes. The average time to unload a patient at the hospital was 4 minutes. All the other times required to simulate the process in Figure 8-2 were also measured.

The model was validated by using travel-time data to derive distributions of the time taken to reach the patient, and time to return to the garage. The distributions from the simulation runs agreed very well with the actual measured data.

A performance measure not yet mentioned is *ambulance utilization,* defined as the percentage of time an ambulance spends on emergency calls. Simulation results were obtained with all the ambulances based at hospital for the following performance measures: ambulance utilization, response time, round-trip time, and time spent waiting for an ambulance to be assigned. Next, a single satellite garage was by the hospital to be close to an area of high call concentration, and all the simulation runs were repeated. A summary of the data with and without the satellite garage is given next.

No significant improvement in service was noted by adding the seventh ambulance to the satellite. The data indicated that with only one ambulance assigned to the satellite, the service was better than can ever be obtained by simply increasing the number of ambulances at the hospital. Because of the location chosen, the satellite is able to improve service to the dense area near which it is placed, but it tends to reduce service to the other sparse area. The data given do not reflect the fact that it requires an average of 17.2 minutes to pick up the patient and return to the hospital after the response time. Therefore, measured against the total response time, the savings produced by the use of satellite are proportionately less.

These simulation results showed that if re-

Total No. Ambulances	No. at Satellite	Av. Response Time, minutes	Time Saved, minutes	Av. Round-Trip Time, minutes	Improvement in Response Time, %	Improvement in Round-Trip, %
7	0	11.8	0	29.0	0	0
7	6	10.0	1.8	27.2	15	6
10	0	11.3	0.5	28.6	—	2
10	6	9.3	2.5	26.5	21	9

sponse time is accepted as a major factor in judging the quality of the emergency service, the use of the satellites gave significant improvement. If the most important factor is getting the patient to a fully equipped emergency ward, the improvements are less impressive. Therefore, the cost of operating the satellite garage must be weighed against achieving the same improvement in service by some other means. For example, Kings County Hospital is not well located with respect to the district that its ambulances service; therefore, redesign of the district might accomplish more than the satellite garages. The ideas that have emerged from this study suggest a more general proposition which could be used as a basis of locating and operating ambulances. This could be stated as: Ambulances should be located where the patients are, without regard to hospital locations, and an ambulance should take a patient to the nearest appropriate treatment center, without regard to the ambulance's home location. The practical implication of this proposition is the separation of the transportation and treatment subsystems of the emergency medical-care system. A system based on this principle is proposed in reference (24). Several simulation runs were conducted to compare the proposed system of dispersed ambulances to all ambulances located at the hospital and the optimal allocation of ambulances between the hospital and one satellite garage. The dispersed system was superior to the other options in that 10 dispersed ambulances will reduce the response time by 30 percent while 10 ambulances distributed 4:6 between hospital and satellite garage will produce only a 22 percent improvement.

A cost-effectiveness evaluation of the alternative ways was done, and the dispersed ambulance configuration was again best, dramatically. Eight dispersed ambulances are as effective as 10 ambulances in a satellite system at about one-fourth the incremental cost per call. Of course, all the simulation data are subject to some error, but the evidence is so compelling that the urban policy maker has excellent information to guide him in making a good choice as to the degree of improvement he desires and the most efficient or least expensive way to achieve that objective.

This work we have reviewed utilized a Monte Carlo simulation technique to determine ambulance station locations. Work is continuing and newer and more complex driving-time models and optimization algorithms are appearing in the literature.

Special Warning Services

In this brief final section we will discuss two special warning services which are recent examples of the use of systems methodology. The first is an earthquake warning system developed under the sponsorship of the California State Department of Water Resources and located on the campus of the University of California at Berkeley. California has a deep concern regarding earthquakes and now maintains 15 seismological stations throughout the state and parts of Nevada. Previously the data from these stations in the form of Richter magnitudes as a measure of earthquake activity were interpreted by personnel at Berkeley and radio announcements kept the public informed. A graphic readout now processes the seismological information in such a way that a continuous representation of the state and progress of earthquakes is instantly available without data interpretation.

The graphic readout is in the form of a large map of the state of California and is dotted with lights corresponding to each of the 15 seismological stations. When the instruments at a given station register ground tremors, the data are transmitted by telephone line to the state capital, where the data are processed and recorded. Next the processed signals are sent to the seismological station at Berkeley and the "tremor flash map." When a signal is received, the appropriate light on the map begins blinking, and a tone coded to identify each station sounds. As a seismic wave moves from one station to the next, the blinking lights mark its path across the state. In the case of small ground disturbances, the flashing lasts only seconds. If the seismological station were to record tremors of a magnitude of Richter 5, the blinking would continue for 3 to 4 minutes. For magnitudes of Richter 6, the map would flash for 6 to 10 minutes. All the earthquake data for the state are processed and encoded in the form of light and sound output for quick and easy interpretation. The distribution of the information to the public and to interested professionals is still being developed.

The output of any warning system, especially if intended for the public at large, will be useful and effective only if the nature and units of the information are known and understood. Often a period of public education is required to achieve understanding and acceptance of a single indicator of a complex process. The Richter magnitudes of the previous example have been broadcast for years and public acceptance is extensive. The simplicity of the Richter scale is a definite asset in the acceptance.

Some cities have attempted to process air-pollution information in such a way as to provide a public warning. Los Angeles and San Francisco have broadcast air pollution and smog information for years. The San Francisco Bay Area Combined Pollutant Index (CPI) is an example of a pollutant index devised for a limited application. The index is designed to inform the public of gross pollutant levels and has no intrinsic scientific meaning. Its primary purpose is to relate a numerical value to the air pollutants *experienced* by the public in the Bay Area. The contaminants included in the combined index are oxidant, carbon monoxide, nitrogen dioxide, and particulates as measured by a coefficient of haze. The equation for the CPI is

$$CPI = 2(O_x) + NO_2 + CO + 10(COH),$$

where
$$O_x = \text{high-hour oxidant, pphm}$$
$$NO_2 = \text{high-hour nitrogen dioxide pphm,}$$
$$CO = \text{high-hour carbon monoxide ppm,}$$
$$COH = \text{largest measured coefficient of haze in the 8–12 A.M. period.}$$

For interpreting the CPI levels, the following arbitrary scale was defined:

0–25	clean air
26–50	light air pollution
51–75	moderate air pollution
76–100	heavy air pollution
101 or greater	severe air pollution

In the CPI equation, oxidant is given a weight of 2

because it is a definitive indicator of the presence of photochemical smog. The coefficient of haze is used to indicate the presence of particulate matter and is given a weight of 10 to adjust its scale to the appropriate visibility-reducing levels relative to the other pollutants.

There are many recorded episodes which show that the affected population can relate the pollution warning index to their own experiences in the atmosphere. In this sense, the seemingly arbitrary aspects of the CPI are justified.

References

(1) J. Chaiken and R. Larson, Methods for Allocating Urban Emergency Units: A Survey, Rept. P-4719, Rand Institute, New York, Oct. 1971.

(2) J. Chaiken and R. Larson, Methods for Allocating Urban Emergency Units, Rept. R-680, Rand Institute, New York, May 1971.

(3) O.W. Wilson, *Police Administration,* 2nd ed., Mc-Graw-Hill Book Company, New York, 1963.

(4) N. Prabho, *Queues and Inventories, A Study of Their Basic Stochastic Processes,* John Wiley & Sons, Inc., New York, 1965.

(5) T. McEwen, Allocation of Patrol Manpower Resources in the Saint Louis Police Department, Vols. I and II, 1966.

(6) R. Larson, Models for the Allocation of Urban Police Patrol Forces, MIT Operations Research Center Tech. Rept. 44, 1969.

(7) G. Carter and E. Ignall, Predicting the Actual Number of Fire-Fighting Units Dispatched, Rand Institute, New York, unpublished report.

(8) J. Kakalik and S. Wildhorn, Aids to Decision-Making in Police Patrol: An Overview of Study Findings, Rept. P-4614, Rand Corporation, Santa Monica, Calif., Mar. 1971.

(9) N. Heller, R. Markland, and J. Brockelmeyer, Partitioning of Police Districts into Optimal Patrol Beats Using a Political Districting Algorithm: Model Design and Validation, presented at the 40th National Meeting of the Operations Research Society of America, Anaheim, Calif., Oct. 27, 1971.

(10) N. Heller and J. McEwen, The Use of an Incident Seriousness Index in the Deployment of Police Patrol Manpower, Rept. NI 71-036G, National Institute of Law Enforcement and Criminal Justice, Jan. 1972.

(11) M. Wolfgang and T. Sellin, *The Measurement of Delinquency,* John Wiley & Sons, Inc., New York, 1964.

(12) A. Tenzer, J. Benton, and C. Teng, Applying the Concepts of Program Budgeting to the New York City Police Department, Rept. RM-5846-NYC, Rand Corporation, Santa Monica, Calif., June 1969.

(13) E. Blum, Urban Fire Protection: Studies of the Operations of the New York City Fire Department, Rept. R-681, Rand Institute, New York, Jan. 1971.

(14) J. Chaiken and J. Rolph, Predicting the Demand for Fire Service, Rept. P-4625, Rand Institute, New York, May 1971.

(15) D. Colner and D. Gilsinn, Fire Service Location—Allocation Models, Rept. 10833, National Bureau of Standards, Apr. 1972.

(16) J. Hogg, The Siting of Fire Stations, *Operational Res. Quart.,* **19**(3), 275–287 (1968).

(17) H. Weitz, A Model for the Simulation of the Fire Services of an Urban Community, *Fire J.,* 48–55 (Jan. 1969).

(18) G. Carter and E. Ignall, A Simulation Model of Fire Department Operations: Design and Preliminary Results, Rept. R-632-NYC, Rand Institute, New York, Dec. 1970.

(19) E. Nilsson and J. Swartz, Jr., Application of Systems Analysis to the Alexandria, Virginia, Fire Department, Rept. 19454, National Bureau of Standards, Feb. 1972.

(20) J. Rockett, Objectives and Pitfalls in the Simulation of Building Fires with a Computer, *Fire Technol.,* **5**(4), 311–322 (1969).

(21) J. Owen, Emergency Services Must Be Reorganized, *The Modern Hospital,* 84–90 (Dec. 1966).

(22) R. Manegold and M. Silver, The Emergency Medical Care System, *J. Am. Med. Assoc.,* **200**(4), 124–218 (1967).

(23) G. Gordon and K. Zelin, A Simulation Study of Emergency Ambulance Service, *Trans. N.Y. Acad. Sci.,* Ser. II, **32**(4), 414–427 (1970).

(24) E. S. Savas, Simulation and Cost-Effectiveness Analysis of New York's Emergency Ambulance Service, *Management Sci.,* **15**(12), B-608–B-618 (1969).

CHAPTER 9

Education

Problems of Large Urban School Districts

In this chapter as in all others we are assessing the impact of systems analysis to date, and reviewing other problem areas which could be approached with systems methodology. If we consider as a significant dimension of the impact of systems analysis, the number of papers and books comprising the bulk of the pertinent literature, clearly the area of education would be the winner. An incredible amount has been written in recent years on the application of the systems approach to problems in education. Broad reading will reveal, however, that the specific problem-solving capability present in all these writings is modest and, instead, other topics are found. There are an incredible number of expository explanations of the systems approach in the education literature. A wide variety of nonmathematical explanations are given, frequently with little attempt at application. This particular effort is in the interest of educating the professionals, because there are a fair number of papers expressing uncertainty about the applicability of the methodology to research in education. Another aspect of the literature is that a large percentage of the writings are in professional journals and represent the efforts of one or, at most, a few people. Solid reports from large, well-funded research teams appear rare. These comments on the systems-analysis aspects of the education literature suggest that there are facets of this problem-solving and research area which are unique. To get some sense of these facets we will study many properties of the large urban school district (*1*).

The attempt to apply systems analysis has

stemmed largely from the high cost of public education. The hope has been that contributions could be made to promote greater efficiency in school operations and ensure more effective use of educational resources. There are, however, intrinsic limitations and barriers to systematic planning in education. Variables such as students, subject matter, teaching methods, and teachers must be related to outputs such as the learning of facts, skills, and attitudes. And this is complicated by the fact that there is no cogent, accepted theory of learning upon which to base models. In addition, the strong political influence of such forces as school boards, parent–teacher associations, teachers' unions, taxpayer groups, and minority-group coalitions have substantial effect on planning decisions.

On the surface it would appear that the systems analyst could easily contribute to problems associated with resource allocation, school bus routing, and student scheduling; however, these areas are also substantially affected by local internal policies. Any systems analysis results would have to be articulated to the community, to the school board, and to other political-action groups. Since districts rely on community financial support, extra pressure is placed on effective conveyance and quality of any recommendations.

In order to be more specific about problem solving in the public education arena, let us look at a critical example. Since about 80 percent of budgets in most school districts consists of salaries, most resource-allocation problems will center around school personnel. In studying the salary problem, we can expect pressure from teacher unions demanding higher salaries, professional organizations demanding greater attention to other factors in salary evaluation besides experience and highest academic degree, economists and other groups in the community asking for differential salary schemes, taxpayer groups demanding control of expenditures, school personnel officers requesting higher salaries for those skilled personnel on which the district has placed a priority, and minority-group coalitions that want more highly qualified teachers for schools in the inner city and loans or incentive plans for teachers who take assignments in the inner city.

The most common method of salary evaluation now used is based upon the fixed-step salary schedule, where the fixed step represents the increment in salary associated with years of experience in the school district. Placement at a step is based on years of experience and amount of formal training. Reference (1) cites these seven shortcomings of the fixed-step approach to salary evaluation.

1. There is no internal consistency of intermediate salaries on the scale.
2. The salary scale does not consider the difficulty of the teaching environment.
3. There is no consideration of the economic supply versus demand mechanisms for teachers with certain specialized skills.
4. Because there are no overlaps in salary between various levels of the school district salary hierarchy, highly qualified teachers have to move into school administration to receive additional salary.
5. The salary factors of experience and education have been shown to be of little value in explaining pupil performance.

6. The allocation of a school district's resources is not compatible with the fixed-step approach. If additional funds become available, they cannot be incorporated in the existing salary schedule in a logical and consistent manner but are usually distributed by an across-the-board salary increase.
7. There is no way to reflect special priorities and objectives of the school district in the fixed-step schedule. For example, some districts might want to attract new graduates by maximizing beginning salaries.

Simultaneous satisfaction of these several requirements is a very complex problem. Previous efforts have shown that multiple regression analysis is not suitable, that linear programming techniques are more promising. Before attempting to apply linear programming methods, a number of familiar issues of the systems approach must be decided. Among the preliminary questions that must be answered are:

1. What are the priorities and objectives of the education the school district is providing? Is it college preparation, social adjustment, citizenship, or what?
2. Having specified the objectives, what are the job functions necessary to attain the objectives?
3. With the job functions known, what training and qualifications are necessary for the people who are going to perform the functions?
4. What hierarchical salary structure will be imposed by the school district in the model? That is, will administrators be paid more than teacher aides, and so on?
5. What characteristics will be assigned to each

of the training and qualification factors in question 3?

The linear programming approach we will utilize calculates the relative weights for each factor so that the ranking of job functions corresponds to the ranking of that function in the school district salary hierarchy. The hierarchy used in this study is based on an individual's responsibility within his job classification. The classification includes superintendents, principals, department heads, teachers, and teacher aides. The factors to be used in the salary evaluation of the salary hierarchy will be determined by contact with the people and groups in the school district who are most directly involved in salary evaluation. The salary factors used in our model are listed next:

1. Learning environment.
2. Subject matter or special skills.
3. Supervisory responsibility.
4. Highest academic degree obtained.
5. Work experience.
6. Special awards and distinctions.
7. College credits completed in addition to degree.
8. In-service units completed.
9. Relative additional workload in the hierarchy.
10. Extra salary for those personnel at the top of each salary classification.

By further negotiation with appropriate district personnel, the difficult task of specifying factor characteristics with their relative weights was accomplished. Several examples follow:

Factor	Characteristic	Relative Weight
Learning environment	Difficult	3
	Medium	2
	Easy	1
Highest academic degree	Ph.D. or Ed.D.	5
	M.A.	4
	M.Ed.	3
	B.A.	2
	A.A.	1
Work experience	12 and over years	7
	10–12 years	6
	8–10 years	5
	6– 8 years	4
	4– 6 years	3
	2– 4 years	2
	0– 2 years	1

The model will derive the values for the factors. For example, if the value of the education factor is derived as \$300, then an M.A. would receive \$1200 for the factor and a B.A. would receive \$600.

The 10 sample factors listed previously will be identified as X_i, where i designates the specific factor. Now that we have the factors, their characteristics, and the relative ratings of the characteristics, it is possible to describe each job function in the organization hierarchy by means of two equations. The first represents the most highly qualified person for the function and hence the highest salary, and the second equation represents the lowest qualifications, hence the lowest salary.

The mathematical form for the equation is

$$\alpha_1 X_1 + \alpha_2 X_2 + \cdots + \alpha_{10} X_{10} = \lambda_j,$$

$$\beta_1 X_1 + \beta_2 X_2 + \cdots + B_{10} X_{10} = \sigma_j,$$

or

$$\sum_{i=1}^{10} \alpha_i X_i = \lambda_j, \tag{9-1}$$

$$\sum_{i=1}^{10} B_i X_i = \sigma_j, \tag{9-2}$$

where

α_i = highest rated characteristics associated with factors appropriate to function j in school district,

B_i = lowest rated characteristics associated with factors appropriate to function j in school district,

X_i = factors associated with job function j,

λ_j = maximum salary to be paid within function j,

σ_j = minimum salary to be paid within function j.

Based on Eq. (9-1) and (9-2), the dollar difference betwen the highest salaries in any adjacent pair of job functions in the salary hierarchy is

$$\lambda_j - \lambda_{j+1} \leq \delta, \tag{9-3}$$

where

δ = specified dollars difference in salary. The lowest salary of the higher classification to the highest salary of the next lower classification in the school district can be adjusted in the model by the following equations:

$$\sigma_j \geq w_1 \lambda_{j+1}, \tag{9-4}$$

$$\sigma_j \leq w_2 \lambda_{j+1}, \tag{9-5}$$

where

λ_{j+1} = highest salary for job function $j + 1$,

σ_j = lowest salary for job function j,

ω_1 = desired minimum percentage overlap in salary between job function j and $j + 1$,

ω_2 = desired maximum percentage overlap in salary between job function j and $j + 1$.

The salary spread in each job function can be expressed as

$$\sigma_j \geq \gamma_1 \lambda_j, \tag{9-6}$$

$$\sigma_j \leq \gamma_2 \lambda_j, \tag{9-7}$$

where

$0 \leq \gamma_1 \leq \gamma_2 \leq 1$,

γ_1 = maximum percentage spread in salary for job j,

γ_2 = minimum percentage spread in salary for job j.

The last restriction to be incorporated in the model is the school district budgetary constraint. Ultimately, it controls the final level of the salary structure through the equation

$$\sum n_{ij} n_{ik} X_i \leq \psi, \tag{9-8}$$

where

n_{ij} = number of school district employees having characteristic k of factor i,

X_i = factor i used in evaluation,

n_{ik} = relative rating given to characteristic k of factor i,

ψ = total amount of funds available for salaries.

If the model is being used for planning purposes, including the acquisition of new personnel, the characteristics of factors 4, 5, 6, 7, and 8 would be unknown. These values would be set at conservatively high figures so that adequate funds would be available for a range of people which the school district might recruit.

The linear programming method requires determination of an objective function based on the established goals and objectives of the school. The model has been developed in such a way that several kinds of functions can be formulated. For example, beginning teacher salaries σ_4 could be maximized if the school district wants to recruit new college graduates. To retain the older experienced teachers, λ_4 should be maximized. To reward inner city teachers, the factor X_1 should be maximized in the objective function.

In the paper by Bruno a numerical example was given in which the factor X_1 was maximized. The linear programming equations were formed and solved for the optimal weights for each factor in the salary scheme. A list of the factors and their calculated values is given next.

X_1	495.83	X_6	1365.63
X_2	100.00	X_7	100.00
X_3	1500.00	X_8	100.00
X_4	100.00	X_9	1138.54
X_5	100.00	X_{10}	161.45

Substituting these values in the model, the following salary ranges were computed.

Superintendents	17,254–22,757
Administrators	13,015–18,618
Department heads	10,377–15,618
Teachers	7,738–12,918
Teacher aides	5,000– 9,118

To determine an individual salary within a range, the summation of the product of the relative weight and the factor weight for each factor the individual possesses is formed.

Clearly, the model could be used in collective bargaining negotiations. Salary demands by a group could be evaluated by the model in a precise, unique, and informative way.

The model is constructed to incorporate both subjective and objective information. It can be used, after negotiation, to reflect the effect of extant political powers, or it can show what the effect of changed attitudes would be. The structure is general and can include as many factors, characteristics, and relative insights as desired to reflect local conditions. The linear-programming format allows a machine solution of a complex calculation involving simultaneous satisfaction of several constraints. A much more detailed description of this work by Bruno is contained in reference (2).

Instructional Systems

An aspect of education that has received much attention from the systems approach is the area of instruction. So much has been written that bibliographies are necessary (3). The intent of systems methodology in this application is to design and manage all the pedagogical components and operations in terms of specific objectives. Any contemplated or existing instruction must be analyzed to determine measurable goals. The analysis of public instruction and the agreement upon the goals of instruction constitutes one of the most critical problems in education (4). Three forces insist that this task be done: the moaning taxpayers, the militant minorities, and the middle-class citizens who are suspicious of innovations such as sex education and sensitivity training. Nationally there is an educational fiscal crisis which may force the construction of cost-benefit models. Without stating the objective of a high school diploma, it is extremely difficult to develop, present, and evaluate a particular course.

The stated educational system objectives can become the basis for decisions about the design and management of the systems of instruction. The contribution of school-building architecture, busses, football games, or computers cannot be determined. At the moment it appears that our learning facilities are, to a certain extent, a collection of habits, stereotypes, and vested interests.

Some of the objections to the systems approach which we examined in Chapter 1 will come into play here. There is a fear that clearly defined objectives of instruction will stifle some of the learner's creative personality, thereby preventing him from moving into areas of unlimited knowledge and widsom. The notion of an "automated" instruction system will be repulsive to many. However, it is fair to say that the organized methods of the systems approach offers what all critics would agree is most desirable, that is, optimum development for each student. Later we will discuss philosophical concerns more carefully, and for the moment will continue to emphasize that regardless of the methodology employed, rigorous and detailed goals and objectives of instruction are necessary. Thompson, one of the leading supporters of the use of the systems approach, offers the following five hypotheses in support of his position (4):

1. The degree to which instruction can be creative tends to vary directly with the degree of emphasis placed upon the objectives of the instruction, and tends to vary inversely with the degree of emphasis placed upon the methodology of the instruction.
2. The degree of ease of designing and operating a system of instruction tends to vary directly with the degree of measurability, the degree of communicability, and the degree of narrowness of the objectives.
3. Some categories of instructional objectives by their nature tend to stimulate uniqueness of personal development, while other categories tend to stimulate standardization of personal development.
4. The degree to which instruction can be individualized tends to vary directly with the degree to which the system of instruction utilizes the principles of the sytems concept and the instrumentation of technology.
5. The average level of significance of the instructional decisions which teachers make tends to vary directly with the amount and degree of significance of the instructional decisions that are assigned to automation.

Thompson defines the goal of the public system of instruction as "an educated person." The educated person is defined as one who interacts with his physical and social environment in a manner that produces personal fulfillment and social benefit which is optimum in terms of his potential. Obviously, many other goals may be defined but let us follow Thompson's example goal to see how it is made specific in terms of instructional design. To obtain more working detail, he breaks down the term "optimum interaction" from above into seven domains where the interaction can occur. The domains are identified as cognitive, emotional, perceptual–motor, social, physical, affective, and aesthetic. While these domains overlap, the important point is that they offer a category in which each type of interaction can be placed. Areas that could fit in these domains but are worthy of special note are:

1. Capacity for divergent thinking.
2. Capacity for problem solving.
3. Personality or character development.
4. Capacity for flexibility in adjusting to new vocations.
5. Capacity for worthy use of leisure time.

Further subdivisions of the seven domains will be exemplified by the breakdown of the first domain—cognitive. The cognitive domain is taken to have six elements: knowledge, comprehension, application, analysis, synthesis, and evaluation. Next, selecting knowledge to reduce further, we note three kinds of knowledge:

1. Knowledge of specifics.
2. Knowledge of ways and means of dealing with specifics.
3. Knowledge of universals and abstractions in the field.

Knowledge of specifics includes knowledge of terminology and knowledge of specific facts. Now pursuing knowledge of terminology, the terminology may refer to mathematics, science, social studies, English, physical education, art, and music. Continuing to the terminology of mathematics as an example, we have:

1. Terminology of sets.
2. Terminology of number and numeration.
3. Terminology of mathematical sentences.
4. Terminology of graphs.
5. Terminology of nonmetric geometry.
6. Terminology of measurements.
7. Terminology of mathematical systems.
8. Terminology of special topics.

The goal analysis is complete when each of the above eight items is expanded in the form of an inventory of the specific terms to be taught.

The example we have reviewed considered only one narrow salient of a potentially very complex analysis. The approach and format is proposed as suitable for an extensive variety of goals of instruction. A great deal of instruction is designed, now based on the attractiveness of the methodology. Examples are closed-circuit television teaching, team teaching, departmentalization, individualized reading, and homogeneous grouping by estimated intelligence. Many of these processes appear and disappear without any real understanding of their specific effects and relative effectiveness having been determined.

Moving now from the issue of goal determination, we want to look at the actual case of a systems approach to the design of an individualized instruction system (5). A science instruction system was developed that incorporated individualized self-paced instruction and laboratory with small-group instruction. The goal of the system was to promote the learning of a set of specific facts which are a part of fundamental biochemistry. Reference (5) is primarily an effectiveness study which compared the new system (B) to the existing system (A), which was a large-group lec-

ture and laboratory approach within a self-contained classroom. System A was a classroom with laboratory stations arranged around its perimeter. The media used consisted of a semiautomatic slide projector, an 8-mm projector and standard laboratory equipment. System B was in a portable classroom and utilized study carrals, each including a desk, study lamp, tape recorder, headphones, and slide viewer; standard laboratory equipment; and a small-group discussion and projection room.

The biochemistry material used for a stimulus was both behavioral and fact-specific in nature. A set of 20 instructional objectives specified the terminal behaviors and the content the students were to acquire during the test period as well as the conditions under which they were to demonstrate their new behavior. The stimulus material was prepared for a teacher-slide presentation in system A, and a tape-slide presentation in system B. The students in B studied as individuals and set their own pace. They received the study material from headphones and operated the slide viewers manually. They were allowed to return to any portion of the tape-slide lectures.

Several examinations were given to test the effectiveness of the two systems, and the findings of the research were that system B was more effective than system A in teaching basic biochemistry. There are a number of detailed ramifications of this evaluation; however, the basic point that system B, an individualized crudely mechanized instructional system with problems of isolation and monotony, was superior in assisting students to achieve the learning objective seems clear.

An educational decision maker might justifiably reject these results, which base a performance measure on existing standardized achievement tests since the relationship between the tests and the instructional-system goals is unknown. Performance on these tests may be determined partly by the desired goals but can be influenced by many other factors irrelevant to these goals. Examples are the quality of the test instrument, testing environment, and state of well-being of the person being tested.

A more sophisticated performance measure is based on "criterion referenced instructional programs" (6). The goals of these programs are stated in terms of observable behaviors that students are expected to exhibit after instruction. The instructional activities are directly related to the achievement of one or more program goals. And then a criterion-referenced test which is deliberately constructed to yield measurements that are directly interpretable in terms of specified performance standards is given. Criterion-referenced tests measure proficiency with respect to a single skill, and there are only two states of proficiency for that skill: mastery or nonmastery at the time of testing. If a student has achieved every objective exactly on schedule, this performance index will equal his school age. Testing would be done throughout the school year whenever an objective had been mastered.

As a final topic in this section on instructional systems, we want to look at means for designing an instructional simulation system. One of the best-known techniques utilizes 13 steps (7). These steps will include some of the points we have emphasized already.

1. Define the instructional problem: Just what is involved? Why was the decision made to do a design; that is, why is the status quo being tampered with? List the information available to help solve the problem, and perform the problem-definition step of the systems approach.

2. Describe the operational educational system: In order to analyze the operational system, the designer must define:

   a. Group of learners for whom the system is being designed.
   b. Manpower available for the system.
   c. Supporting equipment available.
   d. Aspects of procedures, including personnel scheduling, curriculum material, details of courses, and development.
   e. Management features.
   f. Available facilities.
   g. Funds available for both the ongoing and new system.
   h. Philosophy base of the system and the designer.

3. Relating the operational system to the defined problem: This step may cause the designer to redefine or restructure the problem, but compatibility between the defined problem and the available resources must be established.

4. Specific objectives in behavioral terms: Two kinds of behavioral objectives must be specified. Enabling objectives list the knowledge and skills the student must learn to arrive at the terminal state. Terminal objectives specify the behavior that the learner is expected to exhibit after instruction.

5. Generate criterion measures: Both ena-

bling and terminal objectives must be capable of being measured and compared against criterion measures as standards.

6. Determine the appropriateness of simulation: At this point it can be determined if any of the special advantages of the simulation approach are being exploited. If not, because of the cost of simulation, another analytical technique should be used. The seven special advantages that simulation brings to instructional systems are:

    a. When objectives emphasize emotional or attitudinal outcomes, simulations are appropriate.
    b. Simulations integrate affective and cognitive behavior.
    c. Simulations initiate sustained learner activity and motivation.
    d. Simulation allows the instruction system to react to the learner's moves, and the learner can discover the effects of alternative decisions.
    e. Simulation is useful when emphasis is upon incorporation of the behavior desired within the personal domain of the learner.
    f. Simulation allows exercising of behavior under a variety of contexts.
    g. Simulation places a learner in a perceptual frame to sensitize and direct him.

7. Determine the kind of simulation to be used: Three kinds of simulations are used in instructional systems. The first, interpersonal-ascendent simulation refers to the role-playing and decision-making type of simulation, where learners carry a large share of the instructional burden. The second, machine simulation has the instructional burden by media such as slide tapes, films, programmed instruction, and computers. The last category is actually nonsimulation games which bring some of the advantages of simulation games to instruction but do not simulate any social or physical system.

8. Develop specifications for simulation experience: The simulation model is based on a theory of reality. This base must be as good and appropriate as possible to ensure the quality of the simulation.

9. Develop the simulation-system prototype: Most of the difficult conceptual work has been done, and construction of the actual instructional system can begin.

10. Tryout simulation-system prototype: Small groups of learners are taken through the instructional system by the designer. Their reactions are carefully monitered so that system refinement can be done.

11. Modify the instructional-system prototype: Careful evaluation of whether or not system objectives are being met is done, and changes made, if necessary.

12. Conduct field trial: Even though the system may operate successfuly with the designer, the target population must have an opportunity to experience it.

13. Make further modifications to the system based on field trial data: any last defects are removed at this final stage.

PERT in Education

In previous chapters we have discussed the use of PERT techniques in terms of dealing with complex scheduling problems. In this section we will look at an example of the use of PERT in developing an individualized instructional program for elementary school mathematics (8). A team of educators was assigned this task, and the 6-week period available necessitated a careful sequencing and timing of the subtasks. First, without regard to sequencing, the subtasks were identified as:

1. Adopt a system for classifying mathematical objectives.
2. Identify performance objectives.
3. Write objectives in performance terms.
4. Write each objective for the student.
5. Develop a challenge pretest for each objective which is an exact measure of the stated performance with the criterion standard made explicit.
6. Develop a format for the student contract which includes the learner's objective and the challenge test.
7. Develop a format and procedure for making the challenge-test answers available to the student.
8. Investigate, select, and assign instructional resource materials for each peformance objective.
9. Develop a format for a student prescription sheet for each performance objective.
10. Develop a set of five posttests for each performance objective as alternative forms.
11. Investigate and select, or design, a suitable survey test to be used for placement and diagnosis.
12. Develop a course of study map which provides an efficient visual flowchart of progress in performance objectives and makes assignment to new objectives obvious.
13. Develop a cumulative record form for recording progress.
14. Plan a schedule for getting materials to the typist.
15. Develop a teachers' manual to provide a quick and efficient orientation to teachers.

These steps constitute an interesting practical example of a systematic approach to the design of an instructional system. The PERT chart given in Figure 9-1 made it possible to coordinate efforts of the team members and to complete the project in the allocated time. PERT may also be applied to long-range curriculum development, administrative council planning, facilities planning, school board planning, system-wide schedules for opening and closing schools, and individualizing instruction.

Planning-Programming-Budgeting Systems

The use of planning–programming–budgeting systems (PPBS) in education has been discussed extensively in journals and books but as yet is used little in practical settings. Listed next are the traditional budget and the program budget so that their critical differences can be easily seen.

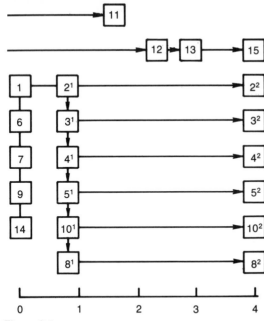

Figure 9-1

Traditional Budget	*Program Budget*
1. A way of matching resources and organizational allotments.	1. A way of matching resources and desired objectives.
2. An annual accounting.	2. A long-range plan.
3. Who gets what?	3. What gets done?
4. Budget change decisions are based on percentage increments.	4. Budget-change decisions are based on anticipated output.
5. Budgets are interpreted as limitations on spending.	5. Budgets are interpreted as current priorities.
6. Investments are allocated as payoffs.	6. Investments are related to payback.
7. The total budget is a summation of incremental changes.	7. The total budget is the best match between available resources and needs.
8. Allotments serve as constraints to decision making.	8. Allotments are determined by decision making.
9. A comptroller's budget.	9. A manager's budget.

Reflecting the differences listed, just the format of a traditional school budget is given:

General Control
    School Board business office and supervision
    Supplies and miscellaneous office expenses
Instruction
    Salaries of teachers and principals
    Textbooks
    Stationary and supplies
    Curriculum program
    Clerical assistance
    Testing program
    Education of exceptional children
Operation of Plant
    Wages of janitors
    Fuel
    Water, light, telephone, and power
Maintenance of Plant
    Repair of buildings
    Repair and replacement of equipment
Auxiliary Agencies
    School libraries, visual education
    Promotion of health
    Transportation
Fixed Charges
    Insurance, rent
Capital Outlays
    Equipment of buildings
    Alteration of buildings

The program budget is not designed to replace the traditional budget but it is designed to provide the manager of the system with a decision-making tool.

The educational program budget is classified by level, function, and purpose, as shown next:

Level	Function	Purpose
I	Instruction Coordinating programs Business programs	Personal safety Health Satisfactory home and community environment Economic satisfaction Satisfactory leisure-time opportunities Transportation, communication location
II	(Under Instruction) Regular Special Continuing	(Under Health) Control of infectious and contagious diseases School nursing Health screening programs School treatment programs (Under Learning) Basic skills Moral and social skills Individual fulfillment skills
III	(Under Elementary) Early childhood Elementary Secondary Vocational	(Under Basic Skills) Physical self-care skills Independent living skills Specific living skills
IV	(Under Elementary) Reading Social studies Mathematics Language arts	(Under Independent Living Skills) Language arts Computational Reasoning

Level	Function	Purpose
	Science Outdoor education	
V	(Under Science) In-class instruction: primary unit In-class instruction: intermediate unit Nature center field trips	

This classification system for a program budget attempts to deal with the difficulty that most school activities satisfy several output objectives. Many other arrangements are given in the literature. McNanama (8) believes that the infrequent usage of PPBS in education is because so little detailed instruction on the use of PPBS is given in the literature. As a result, McNanama gives a detailed worked-out example in his text.

Educational Gaming

The marriage of simulation, gaming, and education creates a rich potential for new learning systems. Abt believes that games constitute a technique that improves student understanding of social studies by means of well-known devices of conditioning through doing analogizing to the students' previous experiences (9). Social studies traditionally includes aspects of history, geography, civics, and economics. Two examples of the basis of gaming from these subjects are as follows. In history, aspects of the Civil War may be learned by identiyfing the players as Loyalists versus Rebels, the objective as to gain support of neutrals, and the resources as coercion and persuasion. In economics, union–management collective bargaining can be studied by role playing the union against the management to gain an increased share of profits.

The design of educational games of the type we have just described can be based on a kind of an elementary design procedure. The parts of the procedure are a systems analysis of the substantive problem, process, or situation to be taught; the design of a model that is a simplified manipulable analog of the process or problem to be taught; the design of a human player simulation of the model; and the refinement of both the original systems analysis and abstract model through repeated test plays of the game.

The analytical portion of the procedure involves limiting the situation or process to be analyzed by applying the educational objectives. All major decision-making entities, their material and information inputs and outputs, and the information and resources exchanged by these decision-making elements are identified. Examples of decision-making elements are producers, consumers, entrepreneurs, and traders in economic models; and tribal leaders in anthropological models. The identification of the flow of information in most political, economic, social, and natural systems is revealed usually to be largely cyclical. This condition defines a "conservative" system in which the total amount of whatever flows through it is a constant quantity.

The game model is formed by identifying the major actors in the process, their interactions, and their decision rules in responding to each others' actions. To determine the criteria used by a cabinet in making its political decisions, we need to know a great deal about their information and resources under a variety of conditions.

To successfully translate the analytical model into a human player game, the model must be translated into a social drama that involves the student's interest and enables him to experiment actively with the consequences of various changes in the system. The game provides dramatic conflict, curiosity over the outcome of uncertain events, and direct emotional expression through role playing. The design must utilize subplots, characters, and events which dramatize the material to be conveyed. Maximum learning is achieved by arranging player objectives, allowable activities, win–lose criteria, and game rules. The active role of the student in a well-designed game appears to invoke greater information-comprehending capacity.

After the educational game is designed, parameter adjustments and design tradeoffs must be made based on a series of test plays. Typical educational game design tradeoffs are realism versus simplification, concentration versus comprehensiveness, and melodramatic motivation versus analytical calm.

Education games may emphasize skill, chance, reality, or fantasy. Games of skill depend on player capability, and they reward achievement, responsibility, and initiative, and discourage laziness. Games of chance dramatize the limitations of skill and chance, humbling the overachievers and encouraging the underachievers. Games of reality, such as Monopoly, simulate real-world operations. They offer the most opportunity for a variety of vicarious realistic experiences beyond the student's direct experiences.

Games of fantasy are intended to release the player from conventional perceptions and inhibitions as in dancing and skiing.

To conclude this brief section on educational gaming, we can say that the games are a combination of systems analysis, dramatic arts, and the educational content to be conveyed. The self-directed learning in games occurs in three ways.

1. Learning facts based on the nature and dynamics of the game.
2. Learning processes which the game simulates.
3. Learning the relative costs and benefits, risks and potential rewards of different strategies of decision making.

Student Registration System

The student registration process in any educational institution is extremely important and traditionally a headache. In the research we are going to cite, which was conducted at the University of Minnesota, the task is to register over 41,000 students enrolled in 21 colleges, having majors in any one of hundreds of departments with thousands of courses (*10*). The student who does not experience some difficulty in registration is unique. The objective of the study was to determine the ideal registration process and then find a way to implement something approaching this ideal. In the light of many of the thorny problems to which the systems approach has been applied, this application seems relatively noncontroversial. Nevertheless, it is a complex problem, and there are a number of alternative solutions.

The objective of a registration process from the viewpoint of a student is to enroll in the desired courses in the least possible time, to spend a minimum amount of time waiting in line, to receive accurate and quick communication of closed classes, and to resolve schedule conflicts with a return trip to the advisor. From the viewpoint of the administration, the goal of registration is to register all the eligible students in the shortest period of time, with a minimum of errors, and for the least possible cost.

A typical current registration procedure utilizes the following steps:

1. The student obtains a registration permit.
2. The proposed schedule is approved by the appropriate person or office.
3. Class registration cards are obtainable by the student for courses with limited enrollment.
4. All the registration material is turned in at a central location.
5. Fees are paid.

The usual problems encountered in this type of registration procedure are a high frequency of closed sections and class conflicts, the need for advisor approval of minor program changes, too much geographic separation between various registration steps, and the long lines and the time required to be spent in them.

A number of existing and proposed solutions to the classic registration problem exist, and these should be considered as alternatives for future work.

1. A registration accomplished by mail.
2. All registration participants and processes located at one geographical point.
3. Submitted desired programs are used by a

computer to build the entire university schedule.

4. On-line terminal registration in which a student communicates his registration via a terminal to the computer.

5. On-line telephone registration, in which students telephone their registration to a central point.

6. Desired schedules are submitted early in a preregistration process and then the college attempts to fit the courses to match the demand of the students.

Each educational institution will have its own unique scaling factors and constraint structure, and the selection of an alternative or hybrid solution will depend on local conditions.

The partial solution chosen by the researchers at Minnesota utilizes an on-line registration system. The on-line aspect is provided by a number of strategically located graphical display readout terminals where the student registers at the terminal while the computer checks the student's course requests for conflicts and closed sections. This kind of an on-line system eliminates many administrative hand-bookkeeping procedures, reduces the number of queues, and will generally decrease the number of employee–student hours involved in the overall registration process. However, the problems not solved by the on-line procedure are the difficulties associated with obtaining the advisor's signature and with communicating the information on closed sections to students and college offices. Therefore, even with the on-line system, student hours in queues will be prohibitive.

The proposed hybrid system uses both on-line computer terminals and the telephone to resolve most of the problems of the classical system. The basic block diagram of the system is shown in Figure 9-2. The steps to be followed by the registering student are:

1. The student obtains his registration forms and referring to the television information on closed sections and courses, forms his schedule.

2. Using an assigned time slot, the student will telephone his schedule to the terminal operator, who will negotiate with the computer to confirm the requested schedule.

3. After the terminal operator successfully enrolls the student, the student is informed by telephone, and a computer printout is mailed to him.

4. Registration fees may be paid by mail or in person.

This proposed hybrid registration scheme was tested by using a simulation. Critical to the simulation was knowledge of the queues that would build up behind the proposed telephone registration system, that is, the arrival rate of calls and the service time distribution. The telephone registration process was divided into four parts and test data were taken. The parts were:

1. Greeting; check on student's identity and correctness of student's data file.

2. Communication of class schedule.

3. Closed-section procedure.

4. Credit check and closing remarks.

The service time for parts 1, 2, and 3 were quite uniform between their extreme points; that is, the distribution was rectangular. Part 4 was a con-

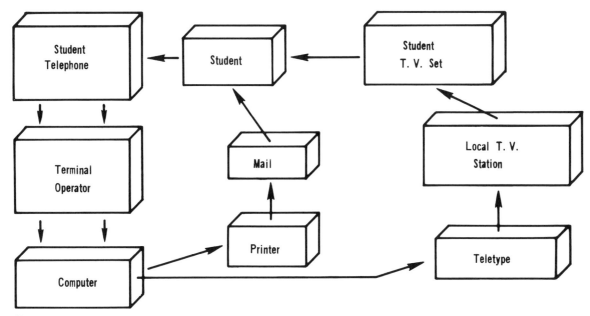

Figure 9-2

stant and taken as 32 seconds. Data on the arrival rate indicated that 80 percent of the eligible population of students for a given registration period is processed in the first one-third of the period. Based on this, an arrival rate distribution composed of two Poisson distributions was postulated. One distribution models the first part of the telephone registration period and the other the last portion of the period.

Using the student data, the model of Figure 9-2, and the stated assumptions, the simulation was run on the computer. On this basis, the following conclusions can be drawn:

1. The average percent utilization of lines and operators is generally low.

2. The average queue length is dependent on average utilization of lines and operators.

These results are very promising, but larger student populations must be run. At the time of this writing further simulations were being done, and considerations of implementing the system on a university-wide basis were in process.

Class Scheduling

In previous chapters and at the beginning of this chapter we have discussed applications of linear programming. For this final case we want to look at a class-scheduling problem which will be modeled with linear programming. The conditions of the problem are:

1. Mathematics classes are to be scheduled for 60 students.
2. Possible group sizes include 60, 30, 15, or 1 student.
3. A total of 1200 minutes of teacher time is available each week.
4. Two minutes of preparation for each minute of large-group teaching is required by each teacher, 1 minute of preparation for each minute of small-group teaching, and no preparation for individual instruction.
5. Each mathematics class or individual instruction will require at least 250 minutes of each student's time, of which at least 5 minutes must be in individual instruction and 20 minutes in small-group instruction.

Weighting factors established by the school administration set the relative value of each type of instruction as 3 for large groups, 5 for medium groups, 8 for small groups, and 40 for individual instruction. The basic issue is to determine the amount of time to be devoted to each type of instruction in order to maximize the total value. If the total value is designated V, then

$$V = 3t_L + 5t_M + 8t_S + 40t_I,$$

where t_L, t_M, t_S, and t_I are the time spent in large, medium, small, and individual groups, respectively. V represents the total value of instruction for each student, and if each term of the equation is multiplied by 60, the total number of students, then V becomes the total value of instruction for all students.

The constraint for teacher time can be formulated as

$$3t_L + 4t_M + 6t_S + 60t_I \leq 1200.$$

Each term in this equation is formed in this way. One large group includes all students and 2 minutes of preparation are required for each minute of instruction; therefore, the teacher time required for large groups will be $3t_L$. A medium group size requires two sections and 1-minute preparation for each minute of instruction, so the teacher time is calculated from $(1 + 1)2t_M$, or $4t_M$, for the second term. Other terms are computed in the same way.

The constraint that each student must spend at least 250 minutes per week in mathematics instruction can be expressed as

$$t_L + t_M + t_S + t_I \geq 250.$$

The requirement of at least 5 minutes of individual instruction and at least 20 minutes of small group instruction is formed as

$$t_I \geq 5,$$

$$t_S \geq 20.$$

In summary, the model of the scheduling problem, formed by a linear programming strategy, is to maximize

$$V = 3t_L + 5t_M + 8t_S + 40t_I$$

subject to the constraints that

$$3t_L + 4t_M + 6t_S + 60t_I \geq 1200,$$

$$t_L + t_M + t_S + t_I \geq 250,$$

$$t_S \geq 20,$$

$$t_I \geq 5,$$

and all times are required to be positive numbers. Solution of the linear programming model on the computer gave the following results. The number of minutes per week for each student was 120 minutes for large-group instruction, 105 minutes for medium-group instruction, 20 minutes for small-group instruction, and 5 minutes for individual instruction. It is an increasing exercise to substitute these values back in the equations of the model and observe that they satisfy all constraints.

Queues in Education

The issue of waiting in lines, or queues, has appeared several times before, and because of the common occurrence of queues in schools another look is appropriate. Generally we believe that if we can apply queuing theory to a problem, the expected length of the line, and the expected waiting time can be computed (*11*). The basic conditions that must be met in order for queuing theory to apply are:

1. Service must be on a "first come, first served" basis.
2. Customers arrive at random time intervals.
3. After arrival, customers wait in line until they are served.

4. The service time of the facility does not speed up or slow down but is independent over time.
5. The facility at which the queue exists has been operating long enough to achieve stability; that is, the number waiting in line and the waiting time become constant.

Queuing theory is based on the assumption that a Poisson distribution describes the time between arrivals in the line. An example of a queue at a single service facility is a film-previewing room used by teachers at a school. The room is used by one teacher at a time for a variable length of time, and the question is whether or not the waiting times in line have become excessive. The decision to build another preview room hinges on the answer to the question. In order to use queuing theory, the investigator needs to know

A = average number of arrivals during a unit of time,

B = average number of customers that can be be served during a unit of time.

Checking the records, he learns that the previewing room has been open 8 hours per day for a school year of 180 days, and there have been 2880 arrivals who used the room for a total of 960 hours. Using these data, A was calculated to be 2 arrivals per hour, and B was 3 customers who can be served in an hour since each customer uses the room an average of one-third of an hour.

Now using well-known expressions from queuing theory a number of results can be obtained.

$P(0)$, the probability of the viewing room being idle at any particular time, is found from

$$P(0) = 1 - \frac{A}{B},$$

and in our example $P(0) = \frac{1}{3}$. This may be interpreted as meaning that, on the average, the facility will be idle one-third of the time. The expected number waiting may be computed from

$$E(w) = \frac{A^2}{B(B - A)},$$

and in our case $1\frac{1}{3}$ people, on the average, will be waiting in line. The expected waiting time is determined from the expression

$$E(T) = \frac{E(w)}{A},$$

and so the average wait for service will be two-thirds of an hour. Clearly, now, the principal has detailed information upon which to base his decision as to whether or not to build an additional facility. Ultimately, however, it is his judgment whether two-thirds of an hour is an excessive waiting time. The analytical efforts have simply provided him with more detailed information upon which to base the judgment; they have not made the decision for him. The problem can be extended by assuming that a second preview room is built, and then calculating the new waiting time. The problem now becomes a multiple service facility and $P(0)$ may be found from

$$P(0) = \left[\frac{(A/B)^F}{F!\left(1 - \frac{A/B}{F}\right)} + 1 + \frac{(A/B)^1}{1!} + \frac{(A/B)^2}{2!} + \ldots + \frac{(A/B)^{F-1}}{(F-1)!} \right]^{-1},$$

where F = the number of facilities. $P(0)$ is computed as $\frac{1}{2}$, meaning that both preview rooms will be idle half of the time. Using expressions modified for the fact there are two facilities, the average waiting time was computed as 1/24 hour.

An interesting adjunct to the preview room problem is to specify that the decision to build will be made based on the economics of the situation. If a teacher's time is valued at $5.00 per hour, with the known number of arrivals and waiting time for the single facility, the cost of waiting is computed as $9600. Assuming that teachers read, grade papers, check lesson plans, and so on while waiting, we can arbitrarily evaluate the cost of waiting at $4800. Making the same computation for two preview rooms the cost of waiting time is found to be $300. The difference in the cost of waiting time under the two conditions is $4500; therefore, if the cost of the second preview room is less than $4500, the decision to build can be made purely on an economic basis. A simplified decision criterion such as this omits several other effects from consideration. For example, the economics criterion does not account for the fact that the existence of the second preview room is likely to increase the demand due to new business generated by the improved facilities.

Input–Output Analysis in Educational Systems

In Chapter 5 we saw how input–output analysis could be used in regional analysis. Its powerful concept of looking at a complex mixture in terms of a few general variables is valuable in educational systems (*11*). Recall that the use of input–output analysis is based upon the following assumptions:

1. Events that occur outside the system are assumed to be known.
2. The effect of change in a level within the system results in proportional change in all other levels in contact with the changing level.

We want to consider the management and planning problem of pupil population at the educational agency level. The levels of the system may be broadly defined as elementary, secondary, and college. The student status for a small urban area is given in the following table:

From Level	To Level: Elementary	Secondary	College	Total
Elementary	7505	0	0	7505
Secondary	0	2500	0	2500
College	0	0	850	850

This table simply indicates where the students are in the system. To determine where the students go when the educational system moves into a new yearly cycle, we form the output transaction table (first table p. 230). The table is self-explanatory; however, there are some interesting bits of information encoded in it. For example, 350 of its student population of 800 are delivered from college to the community each year. Since only 250 students go from the secondary school to the college, there are 100 open places in the college system. Questions must be asked also about the ability of the community to absorb 250 students from high school and 350 students from college. In the next table we add input information, creating a table of input–output transactions (second table p. 230). The input–output transaction table contains such information as:

1. In order for the system to remain stable, it requires an input of 505 students at the elementary level and an additional 100 students at the college level.
2. The secondary level is self-sustaining because the input from the elementary level is sufficiently large.
3. The college must receive 100 pupils who do not come from the secondary school operation in order to maintain its student population. These students will likely leave the area after graduation.
4. The community must provide a student input to the school system of 605 pupils per year to maintain present levels.

The last step in the input–output analysis converts the transactions to a percentage to better show the distribution of students across levels in terms of the prediction of the effect of change (*11*) (third table p. 230). Among the most intriguing results which the analysis has provided is the fact that in order to maintain the college population at

From Level	To Level:			Output to Community	Total
	Elementary	Secondary	College		
Elementary	7000	500	0	5	7505
Secondary	0	2000	250	250	2500
College	0	0	500	350	850

From Level	To Level:			Output to Community	Total
	Elementary	Secondary	College		
Elementary	7000	500	0	5	7505
Secondary	0	2000	250	250	2500
College	0	0	500	350	850
Input From Community	505	0	100	605	
Total	7505	2500	850		10855

From Level	To Level:			Level Output	Total
	Elementary	Secondary	College		
Elementary	93.3	20.0	0	0	7505
Secondary	0	80.0	29.4	10.0	2505
College	0	0	58.8	41.2	850
Level Input	6.7	0	10.8		

850 students, 10.8 percent of that total must come each year from the community. If the appropriate data are available, input–output analysis provides pertinent results for decision making.

Statistical Modeling

In Chapter 2 we suggested that statistics could be the basis of a powerful modeling strategy. Certain educational relationships provide a setting in which statistical specifications are used in an attempt to determine structure and bring order to recurring events. An interesting example is determining the expectation of the future performance of a student. The structure we referred to previously connotes the existence of a source of data and the fact that this knowledge of performance in a past event is related to and has value for predicting the future. If the past performance has a detectable pattern, prediction is simplified. However, the data normally have an error that must be dealt with in order to improve any prediction. Reference (*11*) gives an interesting example of this phenomenon based on the grades of ten students in fifth and sixth grade.

Student	Fifth Grade	Sixth Grade
1	A	B
2	B	A
3	B	B
4	C	C
5	C	B
6	C	C
7	C	C
8	C	C
9	D	D
10	D	C

If the prediction of the sixth-grade performance of these students had been based on the fifth grade, the percent error of prediction would have been very high, 100 percent in the case of students with grade A in the fifth grade. If we change the fifth-grade grades to values according to: A = 4, B = 3, C = 2, D = 1; then assuming a linear relationship between past and future grades we can use statistical methods to compute the specific linear relationship which minimizes error. The well-known method of defining a line that minimizes the deviation about the line we seek is called the *least-squares* method. The basic equation of a straight line is

$$Y = BX + A,$$

where

Y = values of grades to be predicted,
X = current values of grades,
$A = Y$ intercept,
B = slope of the straight line.

Briefly, the computation steps are:

1. Calculate $\sum X$ and $\sum Y$.
2. Compute the average value of X and Y.
3. Form the expressions $\sum X^2$, $\sum Y^2$, $\sum XY$, and compute the slope from

$$B = \frac{N \sum XY \ (\sum X) \ (\sum Y)}{N \sum Y^2 - (\sum Y)^2},$$

where
N = number of grades.

4. Determine the Y intercept from

$$A = Y_{av} - (B \cdot X_{av}).$$

5. With $B = 0.813$ and $A = 0.61$, the line can be drawn as in Figure 9-3. Using the least-squares line, the predicted scores for the sixth grade are:

Fifth Grade	Sixth Grade
4	3.862
3	3.049
2	2.236
1	1.423

The important point to remember about this example is that the statistical model allowed us to detect any order in recurring events so that we could improve the quality of our predictions.

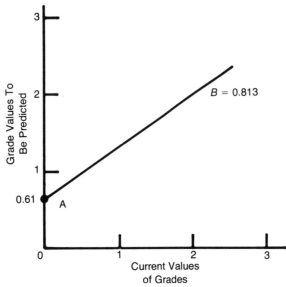

Figure 9-3

Final Comments

As we mentioned in the first section of this chapter, in spite of the slowly emerging applications of systems analysis in education, the literature has many contributors who suggest caution or outright abandonment of the systems approach to educational problems. Perhaps a classic example is Hartley in reference (*12*). His opinion is especially interesting since he himself is a systems analysis practitioner as demonstrated in his book given as reference (*13*). Hartley lists 25 limitations of systems analysis in education:

1. Confusion over terminology.
2. Problems in adapting models.
3. A wisdom lag.
4. Illusions of adequacy by model builders.
5. Inadequate impetus from states.
6. Centralizing bias.
7. Unanticipated increased costs.
8. Goal distortion.
9. Measuring the unmeasurable.
10. Cult of testing.
11. Cult of efficiency.
12. Spread of institutionalism racism.
13. Political barriers.
14. Conventional collective negotiations procedures.
15. Lack of orderliness for data processing.
16. Monumental computer errors.
17. Shortage of trained personnel.
18. Invasion of individual privacy.
19. Organizational strains.
20. Resistance to planned change.
21. Antiquated legislation.

22. Doomed to success.
23. Imagery problems.
24. Defects in analysis.
25. Accelerating social-change rate.

For the reader who had special interest in the objections to the systems given in Chapter 1, it would be especially helpful to read Hartley's paper to determine the details associated with his caveats. As we said at the beginning of this book, if the investigator does not reconcile any philosophical or methodological concerns, little quality work is likely to result. For the reader who would like to know something of a practicing systems analyst's cautions, the final paragraphs of this chapter will review them.

1. Confusion over terminology refers to the wide variation of language and the diverse settings where one finds practitioners of the systems approach. The commonality of outlook or mode of thinking is the thread, and this may not be obvious at all times.

2. Generalized models capable of representing a variety of situations are not common. The adaptation of existing models to other specific problems is often very difficult.

3. "A wisdom lag" is a phrase used by Hartley to identify the breach between the advancement in science and technology and our knowledge of the problems of human relationships.

4. Too often in educational research, models have been promoted as accurate representations of reality when the models may be based on poor data or questionable premises.

5. Small school districts will not utilize systems analysis unless state departments of education totally accept it. The lag in acceptance has occurred, in large part, because of a lack of knowledge and a dearth of demonstrable success.

6. The increased complexity of growing school districts has formalized the lines of decision making in the interest of efficiency. A by-product has been the widening separation between the leaders and the led.

7. The systems approach does not pursue cost reduction as a basic tenet. Any other kinds of gains realized from systems analysis may be accompanied by increased costs.

8. Goal distortion occurs when emphasis is placed on goals that are easily identified and measured while excluding goals that are not easily measured.

9. Many aspects of educational systems appear impossible to measure. And when variables are measured, performance standards may not be available.

10. Hartley fears that testing based on poor instruments, disputable assumptions, incorrectly interpreted data, and purposely manipulated data can offset the advantages of the systems approach.

11. Philosophic objection to systems efficiency is based on concern that preference is given to saving at the expense of accomplishing.

12. The systems approach does not make decisions but provides special results to help the existing process function better. This may be interpreted as a means of perpetuating the status quo and helping to thwart the reform of any existing institutional racism.

13. The economic rationalism of systems analysis may prove unpalatable to the existing power structures and political institutions.

14. An ongoing economic–political power

struggle in education damages the potential for systematic approaches to resource allocation.

15. The absence of a comprehensive theory of educational administration causes a lack of orderliness of content which inhibits the use of systems methodology.

16. The errors associated with the use of digital computers will damage the results of systems analysis.

17. The lack of trained people, adequate financial resources, deficiencies in the training programs of school administrators, and availability of data-processing equipment slows the adoption of systems analysis procedures.

18. Because of the need for extensive data and information for the systems approach, the danger of the invasion of individual privacy must be considered.

19. When organizational dynamics are put in the spotlight by systems procedures, pressures and conflicts may be created.

20. Impersonal systems measures may be incompatible with the human subtleties of education.

21. Existing antiquated legislation will restrict the extent to which programmic priorities can be determined with analytic tools.

22. The validation of the use of new systems concepts may be diluted because the evaluation of an innovative technique may be conducted by the same persons who originally installed the new procedures.

23. The apparent sophistication of the image impacted to systems analysis may work against the acceptance of the value of the new approaches.

24. Any analysis has some defects. Until the defects of the systems approach are understood and dealt with, there will be intrinsic limitations in the systems techniques.

25. The rate of social change is so fast and complex that the associated problems appear to defy systems analysis.

These 25 objections regarding the use of the systems approach in education are given as an example of ancillary issues which may have to be considered in addition to the main systems methodological efforts.

References

(*1*) J. E. Bruno, The Function of Operations Research Specialists in Large Urban School Districts, *IEEE Trans. Systems Sci. Cybernetics,* **SSC-6**(4), (1970).

(*2*) J. E. Bruno, *How to Build Priorities and Objectives into School District Salary Schedules,* A. C. Croft, Swarthmore, Pa., 1970.

(*3*) P. A. Twelker, ed., *Instructional Simulation Systems,* Continuing Education Publications, Corvallis, Ore., 1969.

(*4*) R. B. Thompson, *A Systems Approach to Instruction,* The Shoe String Press, Hamden, Conn., 1971.

(*5*) R. J. Shauelson and M. R. Munger, Individualized Instruction: A Systems Approach, *J. Educ. Res.,* **63**(6), (1970).

(*6*) R. Besel, A Pupil Performance Measure for Criterion-Referenced Instructional Programs, presented at the 40th National Meeting Operations Research Society of America, Anaheim, Calif., Oct. 27, 1971.

(*7*) P. A. Twelker, Designing Simulation Systems, *Educ. Technol.,* 64–70. Oct. 1969.

(*8*) J. McNanama, *Systems Analysis for Effective School Administration,* Parker Publishing Co., Inc., West Nyack, N.Y., 1971.

(*9*) C. B. Abt, Games for Learning, Chap. 3 in *Simulation Games in Learning,* S. S. Boocock and E. O. Schild, eds., Sage Publications, Inc., Beverly Hills, Calif., 1968.

(10) G. M. Andrew et al., University Student Registration Systems Design Using Simulation, *J. Educ. Data Processing,* 217–228 (1970).

(11) R A. VanDueseldorp et al., *Educational Decision-Making Through Operations Research,* Allyn and Bacon, Inc., Boston, 1971.

(12) H. J. Hartley, Limitations of Systems Analysis, *Phi Delta Kappan,* 515–519 (May 1969).

(13) H. J. Hartley, *Educational Planning–Programming–Budgeting: A Systems Approach,* Prentice-Hall, Inc., Englewood Cliffs, N.J., 1968.

CHAPTER 10

Welfare

Of the many urban subsystems that we have considered, the one that appears to be in the deepest state of crisis is welfare. Not only in cities, but almost everywhere, social programs are failing to meet the needs of the poor, the handicapped, the retarded, children, youth, and the aged. It is certain that if present conditions persist, the situation will worsen in time. Contributing factors are population growth, no major reduction in urban poverty, and an increased sensitivity of the poor, aged, and others to their legal entitlements for service. Because of these factors, social and health indices indicate that we are falling behind other developed countries (1).

In one sense what we have described is a paradox since analysis of the welfare problem is aided by the fact that many important issues in welfare are more easily defined then their counterparts in the fields of health, education, and the environment. To set the scene for what we will discuss, let us review the most obvious problems of the welfare system.

1. There are widely varying benefit levels from states, ranging from grossly inadequate minimum incomes for recipients in the poorest states to the creation of perverse incentives for migration from low-cost living areas to high cost–high welfare payment areas.
2. The existing welfare structure creates an incentive for increased dependency. In no states do full-time working male heads of families receive federally aided assistance, and in half of the states even unemployed male heads of families are excluded. The effect is to exclude millions of poor male-headed families, and, in addition, childless couples and unrelated individuals from the federal welfare system.

3. The welfare system contains almost no monetary incentive for maintenance or increase of work effort by welfare recipients. Again the effect is increased dependency.
4. The nonincome eligibility criteria vary widely from state to state and locality to locality and often local administration exercises considerable discretion. The effect is that in one area we can find needy families excluded from benefit because of fiscal and political constraints on caseload growth, and in another area benefits are being received by people many of whom would not be considered poor by commonly held notions of need.
5. In high-welfare-benefit areas, the welfare costs and caseloads are increasing rapidly creating tremendous political pressure and local pressure for assumption by the federal government of an increasing share or even the full amount of the welfare burden.

In order to understand the setting in which welfare systems operate today, it is necessary to discuss the Family Assistance Plan (FAP). The Nixon Administration initially intended FAP as a reform focused on correcting basic structural defects in the system. And even though considerable change in philosophy followed, the initial design of FAP did represent a careful attempt to balance a variety of welfare-related objectives against budgetary and other constraints.

The major aspects of the first form of FAP were:

1. A federally financed minimum income guarantee of $1600 was established for a family of four. (Omitted were the 2½ million poor childless couples and unrelated individuals.)
2. Slightly liberalized provisions for the retention of earnings improved incentives for work effort by welfare recipients.
3. A strong mandatory work requirement was established, and day-care centers and training programs were funded to complement the work requirement.
4. Aged, blind, and disabled welfare recipients under nationwide eligibility standards would receive a rather generous nationwide minimum payment level.

The effect of these early provisions of FAP would be to improve welfare standards for the recipients in the eight poorest Southern states, lessen the discrepancy between the amount of welfare aid paid to female-headed families and that paid to equally needy families headed by able-bodied men, and shift the relative share of the welfare burden away from state government to the federal level.

There have been many changes in FAP since this beginning, but the initial conceptual thrust is clear. FAP has not done much to solve urban problems simply because most of the poor, in general, particularly those not already on welfare, do not live in large urban areas. Those poor who do live in large urban areas are nearly all receiving welfare benefits, and, in addition, when available, food stamps, medical benefits, day care, housing, and other kinds of assistance. Therefore, widely employed definitions of poverty frequently exclude the urban poor, and it is difficult to fault the idea that first priority should go to those who are farthest below any standard. In any case, it appears that the short-run returns to the city from welfare reform will be slight. If there is basic structural welfare reform achieved, over the long term the cities should experience returns in the form of lessened dependency among their populations and, with the establishment of a uniform national welfare standards, a lower concentration of welfare families in the cities.

Decisions in Welfare Systems (*1*)

If the several problems we have just reviewed are to be solved, new directions based on new decisions will have to be forthcoming. The value judgments of many individuals and groups will be involved in decision making for urban social programs. Included are city councils; welfare directors; service workers; citizen pressure groups; recipients and clients; private social agencies such as the Salvation Army, doctors, hospitals, nursing homes, and day-care centers as providers of services; the public media; professional associations; and the business and education community.

The most crucial person in a welfare program is the recipient or client. He decides whether or not to apply for service, or, once having applied, cooperates in services offered. The welfare director, who manages the program, constantly makes operational decisions, and his feelings and aspirations toward its mission and administration are of immediate relevance. The municipal executive and legislative authorities, in whatever form they occur, figure prominantly in the fate, direction, budget, and authority of a welfare program. Their actions are, however, influenced substantially by the voters, the news media, and private social and health agencies. The viewpoints, sympa-

thies, and prejudices of the agency workers influence the quality of the services of the welfare program. And any individual or institutions providing services to a program have to decide whether to make their services available and how much to charge for them.

There are eight general areas where decision making occurs in a welfare system: program objectives, target groups, services, organization of social programs, funding of social programs, budget planning and control, staffing, and facilities and institutions. Considerations in these areas range from mundane operation concerns to the social philosophy of programs, and many decision areas are closely interdependent. Each of these decision-making areas is sufficiently important to warrant a brief discussion.

1. Program objectives: The common goal of social programs is to improve the status of individuals and families; however, specific social programs have widely differing objectives. Vocational rehabilitation and manpower-development programs provide services to make individuals self-sufficient. Programs for aged, blind, and disabled recipients have the objective of helping individuals attain a maximum degree of self-care at home, in institutions, or in foster homes. Other program objectives include strengthening family life, promoting the normal growth and development of children and youth, and protecting abused and neglected children. The onus is on the decision maker to identify both the substance and the limits of program objectives.

2. Target groups: Although the target groups for social programs are usually the poor, disabled, retarded, aged, or children and youth, there is a great deal of overlapping and poor definition of the target groups. An individual may si-

multaneously be involved with the programs of vocational rehabilitation, public assistance, medical services, and, if aged, the Medicare program. Seldom does any group fall into a discrete classification, and decisions must be made regarding identifying the recipient population.

3. Services: The kind, quality, and availability of services to be offered require decision making. The basis of decisions will be the purpose for which the service is intended. Will it be for case finding, education and prevention, diagnosis and treatment, rehabilitation, or for advocacy to represent the interests of the individual family or group to which they belong? Services may be offered for short periods or continuously. Services may be needed sequentially, as, for example, the man who suddenly becomes blind at thirty-five. His sequence of needs will be: intensive counseling to help him deal with his handicap, education to learn the Braille system, special vocational training, and job placement and orientation.

4. Organization of social programs: Some of the options for organization decisions are population group, problem orientation, geography, political jurisdiction, type of service, service objective, source funding, profession, and historical precedence. With this spectrum of possible choices, many organizations are formed based on combinations of options.

5. Funding of social programs: Possible sources of funds include federal government, state government matching, and state/county/city fiscal relationships. Funds may be provided for regular operational services or for special-purpose activities, such as demonstrations, research, evaluation, and training. Federal matching formulas will influence decisions. For example, social services provided under the Social Security Act are matched at a ratio of 75:25 for basic services to families and 50:50 for others.

6. Budget planning and control: A typical welfare quandary is too few dollars and too many claimants. Program results and social needs of one program must be weighed against those of other programs. And intimately involved in this comparison are personal judgment factors, such as confidence in program leadership, knowledge or contact with the program, the public image of the program, and personal or group gains or losses which the decision makers may anticipate.

7. Staffing: The availability of trained manpower is a significant factor in decision making from unskilled to the most esoteric professionally specialized. This difficulty may be so substantial that task analysis and reassignment will be used to substitute nonprofessional workers for scarce professionals.

8. Facilities and institutions: Almost all programs rely on specialized residential and out-of-residential facilities as service centers. Examples are workshops, skilled nursing homes, hospitals, convalescent centers, halfway houses, and neighborhood centers. The permanence of such facilities influence their continued involvement in a program. In spite of the high cost of the initial investment, operating costs, staff costs, equipment replacements, and new construction, opportunities must be found to reconsider their validity in a social program. Changes in research techniques, equipment, and social policy may suggest institutional change.

From the previous discussion of decision makers and the areas where decisions are made, it should be clear that decision making in social programs is highly complex. Other factors not yet mentioned are the fluidity of the programs as new programs are created and existing ones modified; the sensitive social and economic climate in which they function; the dearth of reliable and valid data about program results and effectiveness, program costs, and target groups; the difficulties in laying program plans and setting long-range objectives; and the unavoidable slowness in developing research projects about social problems, producing research findings, and then implementing the research results.

Attitudes of the decision makers are formed based on expediency, whim, personal impression, bias, hearsay, personal gain or loss, and the most objective analysis and evaluation of data and rigorously tested findings.

The range of viewpoints regarding a social program is nearly as extensive as the number of people or groups who think about it. The validation of any decision made in the social-programs area should be based on whether the decision in some way improves the program so that its social purpose to help people is carried out. Even though decision makers may argue that this is their intent, often the decision makers are not of equal importance to the program or to the final decision authority. As a case in point, a rehabilitation director's decision to expand a rehabilitation program is of lesser importance than a mayor's decision to hold the program at its present size. The following criteria appear to meet the test of assuring improvement of program effectiveness at reasonable cost.

1. Social need: The recipient population must need the program or service.
2. Program effectiveness and efficiency: The success of the program or service may be measured in terms of meeting the social need to which it is directed with a minimal expenditure of time, effort, and cost.

3. Program cost: The extent of the economic value of the manpower, funds, and other resources committed to the social program must be considered.
4. Manpower and resources availability: How much manpower, facilities, and institutions and service providers are required by program?
5. Technical feasibility: The state-of-the-art techniques must be developed to the point where their applications match the needs of the program.
6. Harmony among programs: The social program must be compatible with others which are directed to the same individuals and families in a target group.
7. Consistency with social, ethical, and moral standards: Does the social program or service conform to prevailing societal beliefs and practices?
8. Compatibility with constitutional provisions and democratic principles: Programs should be tested for agreement with the fundamental concepts of our society and its democratic foundation.

The advantage of the use of these eight criteria is that they can be the basis of requesting decision makers on opposing sides of an issue to judge the validity of their positions against the criteria.

Objectives of Social Services

A critical ingredient of system analysis, objectives, was mentioned briefly in the previous section. Social services may be defined as those noncash resources made available under public financing for the furtherance of societal goals (2). Any social service activity satisfying this definition is likely to include one or more of the following objectives:

1. The protection of incompetents.
2. The improvement of consumer choice.
3. The enhancement of social functioning.
4. The advance of equal opportunity.
5. The establishment of minimum material adequacy.

The protection of incompetents refers to the inability of some people to make choices in some or all areas of their lives. They are, in effect, wards of the state because of their age, condition, or past behavior. Children protected by adoption or foster homes are examples, as well as parolees, probationers, juvenile offenders, addicts, alcoholics, and the mentally incompetent.

To improve consumer choice, the government takes on the role of an information source in making information available and often in providing the goals and services which individuals would desire if only they could be made aware of the values such goods and services would yield them. Examples are consumer assistance and referral services and government programs in housing, old-age insurance, and individual health care.

Enhancement of social functioning refers to those services which are necessary to regulate individual actions that benefit other citizens. A classic example is immunization from disease, but the individual actions associated with education, recreation, and family counseling bring benefits to other citizens as well.

Equal opportunity for all citizens does not appear to occur under the conditions of the unregulated workings of our economy and politics. Vocational training, job placement, and fair-employment programs promote economic opportunities; legal services, community organization, and ombudsman programs deal with the political aspects of opportunity.

Minimum material adequacy tends to fill in where the first four objectives leave off. There would remain people with earning ability so impaired because of age, disability, or child-care responsibilities that society would want to lift their living standards. Pascal (2) believes that the standard of this disadvantaged group should be lifted, but not necessarily by services. He is opposed to the loss of freedom, dignity, and self-sufficiency involved in the public provision of services in lieu of cash for the purpose of raising living standards. Within the framework of this fifth objective is the concept that those with inadequate incomes ought to be given money with which to choose the goods and services they prefer consuming.

Some social services have combination objectives. Vocational counseling of young workers might be formed under any one of the first four objectives.

There are at least four institutional arrangements to provide the social service which will meet one or more of the stated four objectives.

1. Government production: The government produces and offers a particular social service.
2. Contractor production: An outside institution under contract to the government will offer a given service to an eligible population.

3. Individual benefits: The recipient population through some form of voucher receives the rights to consume a service from the competing offers of service.
4. Cash transfers: Generalized purchasing power is distributed in the form of cash, and the implicit assumption is that "good" things will be done with it.

The second two methods preserve free choice and maintain economic efficiency.

An interesting departure from the existing social welfare practices would be a welfare system of individualized benefits. Some of the apparent advantages are freedom of choice, economic efficiency, increased quality of services because of competition, ease of detection of ineffective services and inadequate programs, reduction in stigmatization of recipients, and the natural rejection of empty or underutilized facilities. On the negative side of individualized benefit delivery is the difficulty of setting standards. There will be difficulty in fostering competition and innovation, and eliminating fraud and exploitation. For example, the G.I. Bill's education and training benefits appear to have worked well in increasing the human productivity of a large fraction of our population. The existing individualized benefit system, Medicare, has fallen short of success not because of the delivery mechanism, but because of the monopoly power and guild loyalty of the relevant professional group. The major point of these last comments is that the individualized benefit recipient is given the opportunity to behave as most of us do—free to make choices, free to withhold patronage, and free to combine for common purposes with like-minded persons.

Systems Approach to Welfare

As a means of being as specific as possible, the following list of items suggests problem areas in welfare systems which are amenable to the systems methodologies (3).

1. In attempting to break down the costs of a service such as institutional care or to identify how the resources of the agency are directed toward a particular service, it is essential to dissect traditional agency structure to learn how the various elements of a service are interrelated and directed toward a particular service objective. In many multiservice agencies it is nearly impossible to determine, without specialized study, what resources are being directed to a particular service objective.
2. The target group and the types of activities an organization may perform must be specified. The relevancy of agency work to the target group must be determined.
3. System objectives must be identified. If the objective of foster family care is given as providing care for a temporary or extended period in an agency-supervised home for children whose parents are not able to care for them adequately, then this is a vague statement of process rather than a precise objective. A real objective would specify the nature of the environment in the foster home which would contribute to satisfying particular needs of the children.
4. In most agencies a number of components of services can be grouped in relation to service objectives. The array of components must be

examined to determine the cost associated with each component. In addition, the identification of alternative ways of reaching objectives and methods of measuring the cost and effectiveness of as many alternatives as possible is necessary.

5. The collection of cost information, program statistics, and personnel information are three important management subsystems of an agency. These subsystems must integrate at the appropriate time and not duplicate each other. Research is needed to establish a common language with common definitions, and an understanding of how the subsystems interrelate.

6. Welfare systems require feedback in order to monitor the system. Through the flow of cost information, program statistics, and personnel information, the output of the system can be measured, controlled, and assessed for its pertinence to service objectives.

Modeling Poverty

The issue of the definition of poverty has been important in the formation of social and welfare programs. Recent literature attempts to define poverty have focused on the individual's or family's permanent level of command over resources. Societal value structure establishes some minimum acceptable level of this permanent income, and a family's condition is measured by the ratio of its permanent income to this threshold (*4*).

Societal efforts to deal with poverty have been to provide various types of training designed to enhance productivity. These programs have been subsidized by the government and the program efficiencies evaluated with cost-benefit analysis. This approach implies that if efficiency is the most desired program characteristic, then mobility from poverty will tend to be restricted to those who are most easily trained and probably least poor. In order to totally eliminate poverty through training, the cost-benefit ratios would be high and the total cost excessive.

The societal attitude that emphasizes training to reduce poverty believes that a dollar income produced by work yields more utility, in a general sense, to nonrecipients than a dollar given to recipients in a welfare program. This reasoning assumes that the loss function for those in poverty is a constant value; however, a person in dire poverty imposes a larger loss on society than an individual just below the poverty line. There is a clear inconsistancy between society's attitude toward work and society's attitude toward poverty. McCall (*4*) suggests a two-variable policy in which both efficiency and distributional effects are evaluated. In this policy the control variables are expenditures on investment in human development and expenditures on income maintenance. Funds are allocated between these two variables so as to minimize the weighted average of the proportion of people in poverty in the long run. The weights are assigned according to society's increasing losses as poverty becomes more profound.

The utilization of any resource-allocation strategy such as this requires the best possible understanding of poverty. A descriptive model has been developed by McCall, and it uses a Markov process to view the movement of individuals among income classes. The simplest version of the Markov model identifies only two income states, poverty and nonpoverty, and the movers between the two states are assumed to obey the Markovian assumption. The parameters required for the model are the proportion of stayers in poverty and nonpoverty and the transition probabilities of those who move between the two states. Techniques have been developed to estimate these parameters based on data from sources such as the Social Security Administration, and they utilize information regarding the proportion of individuals in various income classes and the length of time spent in each class. One version of the poverty model based on Social Security data uses two income states, poverty and nonpoverty; two sex states; five age categories, 16–20, 21–30, 31–40, 41–50, and 51–65; and this is a total of 40 states. Using a model of this type, we can answer such questions as: Is the proportion of stayers in poverty for the group of Negro males between the ages of 31 and 46 significantly different from the same proportion for Caucasian males in the same age group?

The model just described is purely descriptive and it makes no attempt to change the poverty process. The second poverty model we will review interferes with the income dynamics of the poverty process in order to alter the process in the most desirable way. A key assumption for this second poverty model is that there is a fixed budget permit time available for poverty alleviation, and the decision maker can change the poverty process by direct income transfers to the poverty-stricken, by subsidizing the most pertinent training program, or a combination of the two various criteria is used to evaluate the alternatives.

The first version of the model assumes that the transition between states is a Markov process; in one evaluative policy the expected losses imposed on society are minimized, and in the other the steady-state proportion of the people in the poverty state is minimized.

The second version of the model assumes a semi-Markov process is the basis of the poverty-state transitions. A semi-Markov process, one in which the transition time is not constant for each transition but is a random variable with a known probability distribution. The consequence of the semi-Markov assumption is that the transitions from poverty to nonpoverty and the reverse are random variables which are not instantaneous or of fixed duration. In this model the transitions can be affected by the decision maker, who can choose any feasible combination of investment in human development and direct income transfers. Another evaluation criterion is to minimize the weighted average of the proportion in poverty, where the lowest income classes are assigned the largest weights. These models just described depend on extensive suitable data being available and represent an approach to dealing with welfare problems which is not likely to be utilized for some time.

Systems Approach to Employment Problems

At the time of this writing the youth component of the national labor force is in excess of 22 percent of the working population. It is predicted that in the next several years the number of people 20 to 24 years old in the job market will grow at a rate 2½ times that of the labor force as a whole. Unemployment rates among youth have remained persistently high, and the young people now entering the labor force face this gloomy prospect complicated by the presence of racial discrimination, poverty, insufficient education, and the pressures of an urbanized society. A variety of statistics are available to substantiate the depth and extent of the social problem represented by low-income youth in cities. There is national concern about the job problem of youth, yet it is not clear what the nature of the causal connection is. At this time the impact on the total labor force is not serious; the inequality of material conditions between the youth and adult population is not a matter of pressing interest; and the relationship between youth employment disadvantage and the delinquency and disorder ascribed to youth is not obvious. The national concern may stem simply from the general acceptance of greater equality of opportunity as a national goal. Some inequalities exist because of differences in employment performance due to different values placed on goods and leisure, and, in general, on the many nonpecuniary aspects of work. Government intervention is not appropriate in these cases, nor is there much interest in ironing out inequalities due to intrinsic differences in ability among people. The public objective with respect to the economic performance of youth can most simply be expressed as an attempt to increase the degree to which a youth's expected lifetime economic prospects are a function of his tastes, preferences, and innate abilities, and to decrease the degree to which his economic future is dependent on the color and the income of his parents. The issue, then, is economic outcome for our youth. Economic criteria can be measured and correlated with other outcomes such as political power, so-

cial acceptance, and psychological health. Our model of youth employment prospects emphasizes economic factors but allows inferences about behavioral aspects.

The public concern about youth employment has brought into existence a broad variety of public programs, ranging from compensatory education through antidelinquency and antidropout, to skill training and job-placement programs. Program design and adequacy is dependent upon understanding the complex, dynamic interrelationships that underlie youth behavior and youth opportunities, and this understanding does not yet exist.

The model of youth employment developed by Carroll and Pascal (5) consists of a set of simultaneous equations. The purpose of the model is to predict the economic prospects for an individual on the basis of his experiences, tastes, abilities, perceptions, and opportunities.

The first equation in the model allows the evaluation of the present value of lifetime earnings at time t:

$$L_t = \sum_{j=0}^{\infty} E_{t+j} (1+i)_{-j}, \qquad (10\text{-}1)$$

where
L_t = discounted value of expected lifetime earnings at time t,
E_t = expected earnings over any period t,
i = discount rate.

The equation accounts for not only the amount of income an individual will earn in his lifetime, but also the time profile of those earnings. Therefore, the equation models the precise time at which earnings occur, and this is important.

In order to evaluate Eq. (10-1), we need to determine E_t, the expected earnings of an individual at each point in his life. The second equation of the model is

$$E_t = \sum_i T(A_t = A^i)w_t^i, \qquad (10\text{-}2)$$

where

   T = "truth value," equals 1 if the statement in parentheses is true; 0 otherwise,

   A_t = activity in which the individual is actually involved during period t,

   A^i = ith activity,

   i = five mutually exclusive activities, including attending school, working, attending a vocational training program, serving in the armed forces, or "other,"

   w_t^i = wage rate associated with each of the possible activities.

The intent of Eq. (10-2) is to encode the fact that we know what earnings an individual will receive in any given period if we know in what activity he is involved during that period and what wages he receives as a consequence of that choice.

The third equation describes the process by which youth choose among the alternative activities:

$$A_t = f_a\,(F_t^\sigma, F_t^\omega, F_t^\tau, F_t^\alpha, F_t^\psi, S_t, C, t) \qquad (10\text{-}3)$$

where

   $F_t^i(i = \sigma, \omega, \tau, \alpha, \psi)$ = anticipated future returns from each of the activities,

   S_t = state in which the person is found at the beginning of period t; it is a vector of the individual's acquired attributes, such as the number of years of school completed, the number of years of work experience, possession of a high school diploma, completion of a training program, possession of a felony conviction record, and so on,

   C = characteristics of the person, such as race, IQ, preferences, family life, and so on.

Thus Eq. (10-3) describes the activity chosen as a function of the several variables.

The relation between the activity chosen by an individual, A_t, and the activity which actually occurs, A_{ts}, appears to be random from the point of view of the individual. We represent this as an auxiliary equation to (10-3):

$$\text{Prob}(A_t = A^i) = f_b^i(A_t, C, S_t, \ldots), \qquad (10\text{-}3a)$$

where

   $i = \sigma, \omega, \tau, \alpha, \psi$. Therefore, five equations are required for Eq. (10-3a).

For Eq. (10-3) we need to evaluate F_t, and this is

$$F_t^i = P_t^i + rF_{t+1}^i \qquad (10\text{-}4)$$

where

   P_t^i = earnings that are expected in period t if activity i is chosen at time T,

r = psychological discount factor that the youth applies to future income,

F^1_{t+1} = subjective aggregate of the individual's anticipations for all periods beyond the present.

Therefore, the total equation for F^i_t expresses the discounted sum of the youth's earnings expectations for each period in the future. Data, of course, will not be available, but aggregations of present and future expectations could be used.

P^i_r, the expected earnings for activity i in period t, can be determined by these expressions:

$$P^\sigma_t = W^\sigma_0, \tag{10-5a}$$

$$P^\omega_t = f_b(W^\omega_t, T_t, L_t, C), \tag{10-5b}$$

$$P^\tau_t = W^\tau_0, \tag{10-5c}$$

$$P^\alpha_t = W^\alpha_0, \tag{10-5d}$$

$$P^\psi_t = f_e(\cdots). \tag{10-5e}$$

The first, third, and fourth earning returns are easy to determine because they are fixed by public policy. Equation (10-5b) determines the individual's perceived returns from work as a function of the actual wage rate W^ω_t, the cost of transportation t_t, local labor-market conditions L_t, and the individual's personal characteristics C. Equation (10-5e) is not specified in the model since it designates any other activity not included.

To evaluate Eq. (10-5b) we need an expression for W^ω_t, the demand for the individual's services on the labor market:

$$W^\omega_t = f_0(S_t, C, G_t, D_t), \tag{10-6}$$

where

G_t = general economic conditions,

D_t = prices of other factors of production.

And finally, the state in which an individual finds himself at the beginning of period t is

$$S_t = f(S_{t-1}, A_{t-1}), \qquad t > t_0, \tag{10-7}$$

where

$$S_{t_0} = f(C).$$

Equation (10-7) symbolically asserts that the state in which an individual is found at the beginning of period t depends upon the state in which he was found at the beginning of the previous period and his activity during the previous period. S_{t_0} accounts for the starting point of an individual.

Perhaps the most interesting aspect of the employment model is the choice of variables and parameters used to build the model. There is no systematic way to do this, but, once chosen, the model is woven with them. The simultaneous equations of this model satisfy common sense in the way they are put together; however, the data to support them would be difficult to collect in many cases. If we assume that the knowledge accumulated by social scientists can be used to generate measures of the variables C, G_t, D_t, and L_t, then given C we can predict S_{t_0} and A_{t_0}. Continuing these steps of operations, we can generate the entire sequence of activities and earnings which are relevant to the youth over his lifetime, and this is the best indicator of economic opportunity in any long-run sense. As an example of the use of the model, we could ask the basic question: Does the relationship between lifetime employability and the acquisition of work habits, skills, and experience justify a public concern for youth? A way to answer the question would be to compare the lifetime activity path of a person who experienced unemployment at $T(A_t^1 = \psi)$, with a similar youth who had no employment ($A_t^2 = \omega$, or α, or τ). As the magnitude of the quantity $(E_t^1 - E_t^2)$ increases, the more justified an intervention which will reduce the probability that $A_t^1 = \psi$.

The model is conceptually satisfying because it facilitates the comparison of a number of phenomena which have been offered as explanations of the unsatisfactory economic futures that confront poor youth. For example, cultural-deprivation explanations of poverty imply that social forces applied through the family, neighborhood, and peer group so constrict experiences and attitudes of youth as to substantially inhibit possibilities for emergence from poverty. This issue could be studied in the model by finding those components of C which have an important influence on the choice of activity. The activity-choice equation (10-3) can detect the conditions which imply that poverty induces usual requirements such as early gratification, high values on leisure, and a tolerance for low-status occupations. Another explanation which can be checked out is that low income or slum residency establishes an obstacle to the receipt of information necessary for the making of optimal, future-ori-

ented decisions. Other hypotheses that can be investigated include measuring a skin color or social class to trace the implications for wage perceptions, activity choice, and ultimately for lifetime earnings.

The model could be used to test the efficiency of programs designed to remedy inequities in economic opportunity; to test the ways in which a training program affects the ultimate objective; to determine the effect of a fair employment law; to observe the consequences of less-direct attacks on the effects of prejudice; and to explore policy options, such as the easing of minimum-wage laws and Armed Forces induction standards, the raising of family income status through programs of income maintenance, the provision of more accurate market information to job seekers, and the denial of information to employers on aspects of the police records of job applicants.

Evaluation of Social Service Delivery Systems

We have been discussing in this chapter how systems analysis has and will continue to contribute to welfare reform and the establishment of effective social service delivery systems. The separate system of social services lies beyond welfare payments to assist all vulnerable individuals threatened by societal or individual barriers to economic self-sufficiency and independent living. A computer model has been developed for a community-based social service center where the function is to open the doors for clients to the range of available resources and services in a community (6). The center is one of several that has been federally funded to test different types of community-based social service centers. The center upon which the computer model is based is located centrally and represents a single location where any resident can go to get help with obtaining public assistance, employment, housing, child welfare services, legal aid, family planning, medical treatment, home and financial management counseling, assistance with parental and marital problems, and so on. The center can act both as an advocate and resource mobilization point for all its clients in the community. The objective of these demonstration centers is to show that there will be a marked improvement in client adjustment, and a positive movement out of dependency status by clients. The computer model was developed to evaluate the extent to which these objectives are achieved. The model entitled "The Case Service Effectiveness System" (CASE) is applied to a random sample of the case load at the community social services center.

The model was developed to measure center effectiveness based not upon the quantity of services provided or numbers of people saved, but upon the extent to which service provision led to actual problem solving. This approach implied that regardless of the skills with which services are provided, systems effectiveness requires that problem solving result. Effectiveness depends on follow-through and availability and appropriateness of community resources applied to the actual problem solution.

The CASE model is organized around four major areas:

1. Each client served is identified by suitable information and socioeconomic characteristics.
2. The specific social service objectives are set.
3. Service activities and community resources are utilized to meet service objectives.
4. The client movement and the viability of the movement toward objectives is assessed.

Data acquisition for these areas was based on a comprehensive list of service objectives, a list of activities undertaken to obtain the objectives, and an inventory of the range of community resources to be utilized in problem solving.

The CASE model can be used in these ways:

1. The program areas in which the center is most effective can be determined.
2. Specific service activities can be related to objective achievement.
3. Areas of service demand over time and the types of clients with whom social services are most effective can be determined.
4. The center's operations can be optimized by using the data related to the cost of service provision.

The first part of the CASE model is a computer compilation of selected socioeconomic characteristics of the people served by the center. The second part of the model illustrates the distribution of service objectives in program areas over a specific time period. The progress toward satisfying objectives is evaluated based on the following scale:

0 Premature termination of objective
1 No movement
2 Some progress; objective achievement not in sight
3 Considerable progress; objective achievement in sight
4 Objective achievement

A subjective assessment of the client's movement toward each service objective is made and the validity of the assessments is checked by client interviews.

When each social service objective is fully achieved, a social service worker determines for evaluation purposes whether

1. The client is likely to be dependent on the center in the future for similar problems.
2. The client has progressed toward solving similar problems in the future.
3. Don't know/not applicable.

On the basis of this second part of the model, it is possible to pinpoint areas of service weakness based on the client movement scale and make suggestions for service improvement.

The third part of the model prints out the range and extent to which community resources are being utilized in case problem solving. Weak links in the community service system can be identified by this means.

In summary, the CASE model for evaluating the effectiveness of community-based social service centers was extensive data, subjective scales, and computer organization to accomplish its purposes. It is hoped that CASE data will be used to forecast the future demand for services and identify gaps in total community capability to effectively help the disadvantaged.

Child Care Centers

A social services program under consideration is a federally funded child-care program. Child care is a national priority and although child-care centers are likely to become a significant part of our society in this decade, many studies of the possibilities are being made. Galvin has considered child-care centers from the perspective of the cost effectiveness of the program to the federal government (7).

The progress toward the development of a national child-care program has been steady since the Social Security Act was amended in 1967. Success hinges upon the completion of suitable research and a favorable political environment. A number of important economic and technical considerations need further development before child-care centers can take the position in our society that they have in other countries. Legislators are most frequently turned away from support of a federally sponsored program by financial factors. Although we will discuss economic aspects primarily, the social and educational benefits remain as the foundation of the program. Child-care centers offer a real hope for breaking the generations of poverty of so many families.

It has been shown that providing a job for an unemployed person contributes $10,000 to the gross national product, creates $241 in income tax revenue, $36 in state sales tax, and increases purchasing power for goods and services by $3400. However, to realize these figures, to allow a woman to work and free her from welfare, a substantial price tag for the child-care center must be met. One of the greatest blocks to significant federal legislation is that many child-care proponents speak of a $10 billion price tag, that is, 5 million children at $2000 per year per child. And this figure does not include construction and developmental cost.

Galvin has studied the problem of the $10 billion cost and has developed a counterposition to the issue of economic burden. He illustrates the gain in revenue by assuming that a single parent currently on welfare has one preschool age child and that she obtains a minimum-wage job ($1.65 per hour in California). Assuming that she does not qualify for a reduced welfare payment, her income after employment will be $3432; she will pay $238 in federal income taxes, $357 in Social Security tax, and the federal contribution to her welfare grant, $1332, will no longer have to be paid. The gross economic gain to the federal government is $1927, and this would more than pay for the cost of caring for this parent's child in a child-care center.

Galvin's case is strengthened further when the inefficiency of the disbursement of federal revenues for welfare use is considered. In addition to the direct economic gains of the example, there are secondary gains such as the increased purchasing power, increased gross national product, and a general stimulus to business and industry. There would also be a gain to state and local governments in the form of income tax and sales tax.

The revenue gain of the previous example has been oversimplified, and a closer look is necessary. For example, in addition to a lack of adequate child care, there are many other factors which prevent the removing of individuals from welfare rolls. Included are training, housing, medical, transportation, and discrimination. The gross federal economic gain in revenue is also affected by the employment of a former unemployed person who was not previously receiving welfare payments. In this case Galvin's computations show that if the mother were to obtain a job that pays at least $2.69 per hour, she would create enough federal revenue to support her child in a child-care center.

Other children up to about age nine would also require child care; however, the costs decrease substantially. Galvin estimates that if a mother begins employment when her child is 6 months old, over the next 8 years the cost of child care will be $11,600, but she will return over $13,000 to the federal government in tax revenue.

The whole point of Galvin's analysis was to show that the financial aspects of a federal child-care program are positive rather than negative.

Other important aspects not discussed are:

1. The difficulty of changing national attitudes regarding the role of a mother.
2. The source and availability of employment for some 5 million potential job seekers.
3. Most child-care centers require the parents to pay a portion of the cost in the form of fees or tuition.
4. The existence of a federal child-care program would create more than 1 million new jobs in early childhood education, health, and social work.
5. The balance of enrollment among rich, poor, and various ethnic backgrounds will be a difficult task to accomplish.
6. In order to maintain the personal contact so

essential in a child-care center, a modest-sized center—50 to 90 children—should be maintained.

Clearly, a multidisciplinary approach accomplished by the systems analyst, educator, health official, social worker, and child psychologist is needed to make progress in this field.

Welfare Maximization

If we examine the current impact of systems analysis on public systems we find that the impact is modest. The best design of a public system has not been defined accurately thus far. The systems analyst can offer only the elements of a social welfare function to the public and its decision makers. The social welfare function is defined as the unique function which expresses the total welfare of the public as a function of all the goods and services available. In this sense, the social welfare function is the utility function, as described in Chapter 2, of society as a whole (8). This is a much broader theoretical concept than anything discussed thus far, but it is the best we have to offer in the area of welfare. How the welfare of society and the utility of individuals relate is not known. A possible way of looking at it is that society is composed of many combinations of two-person communities whose conjectured utility function is given Figure 10-1. The figure lends credence to the point made in Chapter 1 that in a social system when the status of one person or group is improved, it is at the cost of the status of another person or group.

Figure 10-1 is a generalized social welfare function for an individualistic society. A commun-

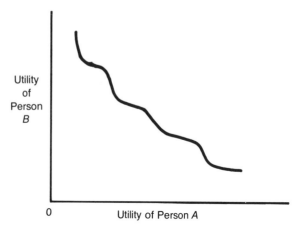

Figure 10-1

al society might have a social welfare function as depicted in Figure 10-2.

Both Figures 10-1 and 10-2 indicate only the generalized curve shapes, and a more explicit shape would depend on the moral and ethical

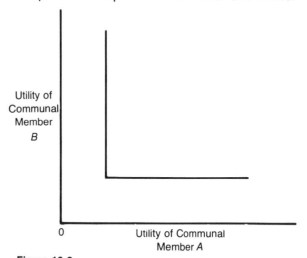

Figure 10-2

foundations of a society. It seems reasonable to assume that the social welfare function of the United States would lie somewhere between the two extremes of the individualistic society and the communal society. Many of our social welfare programs emphasize the needs of disadvantaged populations; therefore, their social welfare functions are actually closer to Figure 10-1, that is, generally treating people equally but sensitive to special needs.

Using the techniques of psychophysics, public referenda, polls, and surveys, one might try to determine the public's utility function, but it has not been done. In the absence of this information, the analyst must confront the process of examining the relative significance of the social objectives.

A group consensus on community preferences can be achieved by compromises between members of the group. For example, voters can trade votes to gain support on issues of concern while promising to vote for someone else's favorite project. By studying this phenomenon, the systems analyst can determine the information necessary for evaluation. The expected utility of their votes can be maximized by the members of a community by casting it in the way that does them the most good. Again this kind of information could be used to get some idea of the nature of the social welfare function.

Other studies have shown that it is not possible to determine social utility by comparing each person's utility and transforming it into an appropriate common denominator. The measurement of personal utility based on current assets and moral and aesthetic judgments is conceptually feasible but extremely difficult. No way is known to combine individual utilities to a common base. For example, if one person decides to be relatively poor rather than endure the stress of pursuing success, how can his utility be combined with that of an individual who seeks financial success?

Any discussion of social welfare, specific or general, must inevitably face the basic issue of public policy: Who pays and who benefits? Those particular populations must be identified for each situation, and the task is difficult. Early thinking on this subject is exemplified by the nineteenth-century economist, Pareto, who postulated that social welfare increases whenever one or more individuals achieve a higher utility without any other individuals becoming less satisfied. Later studies took the position that Pareto's postulate defined only a subset of the circumstances under which social welfare will improve. An alternative condition, mentioned previously, says that total utility increases where someone's utility decreases. As an example, take the situation where 90 percent of the possesions of a millionaire were confiscated and given to 1000 people in poverty status. It is conceivable that although the millionaire would be unhappy, the welfare of society as a whole might be increased.

An extension of the conditions just discussed is the compensation criterion, which states that any function which fully compensates each individual for any losses caused by the function and which increases the utility of some people must represent an increase in general welfare. This principle is in accord with the standards of equity which prevail in the United States and societies where private property and property rights are protected by law.

We have said that the systems analyst cannot determine the utilities of individuals or define a social welfare function with accuracy. Further, it is clear that economic criteria cannot be the basis of maximum welfare. Much study needs to be done on the many details and ramifications of the following statement: The value of a social program depends not only on people's individual values, but also on who receives the benefits, how the benefits are distributed, and who incurs the costs.

References

(1) A. Spindler, Decision Making: The Target of Systems Science in Social Programs, presented at the Joint National Conference on Major Systems, Anaheim, Calif., Oct. 29, 1971.

(2) A. H. Pascal, New Departures in Social Services, Rept. P-4079, Rand Corporation, Santa Monica, Calif., Apr. 1969.

(3) R. Elkin, The Systems Approach to Defining Welfare Programs, *Child Welfare,* **46**(2), 72–77, 1970.

(4) J. J. McCall, An Analysis of Poverty: A Suggested Methodology, Rept. RM-5739-OEO, Rand Corporation, Santa Monica, Calif., Oct. 1968.

(5) S. J. Carroll and A. H. Pascal, A Systems Analytic Approach to the Employment Problems of Disadvantaged Youth, Rept. P-4045, Rand Corporation, Santa Monica, Calif., Mar. 1969.

(6) R. M. Clinkscale and M. P. Rymer, An Effectiveness Evaluation Model for Community-based Social Service Delivery Systems, presented at the Joint National Conference on Major Systems, Anaheim, Calif., Oct. 29, 1971.

(7) D. M. Galvin, Child Care Centers—An Analysis Overview, presented at the 40th National Meeting of the Operations Research Society of America, Anaheim, Calif., Oct. 28, 1971.

(8) R. de Neufville and J. H. Stafford, *Systems Analysis for Engineers and Managers,* McGraw-Hill Book Company, New York, 1971.

Part IV
The Promise of the
Systems Approach

CHAPTER 11

The Future

Societal Problems

In the preceding chapters of this book one of the basic messages intended for the reader has been that systems analysis in the very nature of its dependence on quantitative methods has been most successful in fields that have a large technical content. Successes have been extremely modest in any problem area which has a substantial emotional or psychological content. Although short-range goals for urban improvement appear fairly well defined for each of the specific subsystems, the mechanism and rationale for reconciling conflicts among competing subsystems does not yet exist. Definitions of overall urban system goals are not clear, and there is not adequate understanding of how to plan and operate the city as a total system. And further, how the urban problems fit in societal problems in general is not known. As a means of being more specific about these issues, this section will consider a systems approach to societal problems.

The power and appeal of systems analysis lies in its problem-solving ability. When the concern is societal problems, the first issue is the classification of the societal problems into an ordered list which tells us what to do and which problem to work on first. In other words, what are our national priorities for societal problem solving? The literature contains many examples of such lists. In six recent attempts to classify the entire spectrum of societal problems, 6 to 20 categories were used. Only a few categories were common to two or more lists, and some categories overlap with those in other sources, but in the main, each list is unique and appears to be relevant. For example, the Department of Health, Education, and Welfare proposes the following list (2):

1. Health and illness.
2. Social mobility.
3. Physical environment.
4. Income and property.
5. Public order and safety.
6. Participation and alienation.
7. Learning, science, and art.

Clearly, the categories are broad. Another source classifies societal problems in the following way (3):

1. Divisions in U.S. society.
2. International affairs.
3. Education.
4. Urban area.
5. Law and order.
6. Science, technology, management of change.
7. Economy.
8. Resources.
9. Value.
10. Population.
11. Political economic power.
12. Business–government relations.
13. Information collection, processing, distribution.
14. Political structure and parties.
15. Food.
16. Transportation.
17. Business and organized labor.
18. Behavior control.
19. Genetic control.
20. Leisure.

At least, in this list, urban problems appear explicitly.

The following table (1) demonstrates that many of our present societal problems have some of their roots in a technical or scientific advancement.

Successes	*Resulting Problems*
1. Prolonging the life span	1. Overpopulation; health and custodial care for the aged
2. Machine replacement	2. Exacerbated unemployment, and accelerated urbanization through displacement of farm workers
3. Efficient production systems	3. Dehumanization of work evaluation through over-emphasis on efficiency criteria
4. Advances in transportation	4. Increasing air, noise, and land pollution and urbanization; information overload
5. Nuclear and biological weapons for national defense	5. Hazards of mass destruction
6. Satisfaction of basic needs and consequent upward movement	6. Rebellion against non-meaningful work; unrest among students
7. Growth in the power of systemized knowledge	7. Threats to privacy and freedom through surveillance technology
8. Material affluence, especially in developed nations	8. Increasing per capita energy and goods consumption, leading to pollution and resource depletion

There are many controversial points in the previous table; however, it seems clear that problems of that ilk cannot be solved or managed by conventional strategies, especially if it is a piecemeal approach. As stated in Chapter 1, these societal problems have no solutions in the conventional mathematical sense. As in the case of medical problems, "solutions" come in the form of remedies of varying effectiveness.

Any remedy for a societal problem is actually a process with three basic components:

1. The establishment of a plan, strategy, and schedule.
2. The people who define and understand the problem and participate or concur in the remedy.
3. The allocation of the physical and economic resources needed.

A class of normative problems usually involves serious or irresolvable conflicts of values, cultural norms, objectives, or priorities. Examples are:

1. Bureaucratic behavior, where persons in their institutional roles treat other people as though they are objects or things.
2. The tendency for individuals acting as agents of a larger body to seek to maximize the short-term good of the group whom they represent rather than accepting a larger sense of social responsibility.
3. The viewpoint that any technolgoical breakthrough that can be profitably developed should be—and this leads to problems of market saturation, unregulated growth, and resource depletion.
4. The concept of legitimacy, which is defined as widely held beliefs that present institutions are the most appropriate ones for society and that they reasonably reflect individually held preferences and views. Examples of three such beliefs which are eroding now are:
   a. The belief in a common value system from which we can set national objectives and priorities.
   b. The belief in the state's monopoly of coercive force with which it can repress behavior of dissidents.
   c. The belief in the honesty of the administrators of commonly held resources.

To conclude this brief discussion of the taxonomy of societal problems, it is appropriate to look at a classification of remedies, all of which may be achieved by the systems approach. The four types of prototypical remedies are systemic, preventive, ameliorative, and compensatory. As an example, the following are some remedies for reducing the suffering caused by highway traffic accidents:

1. Systemic: Design safer better forms of public transit.
2. Preventive: Require treatment for convicted drunk drivers.
3. Ameliorative: Reduce injuries by improving the crash worthiness of automobiles.
4. Compensatory: Self-insurance is required in order to reduce delays in remuneration of accident victims.

This approach to a strategy of remedies for societal problem solving has great promise.

Library Models

To a large extent, the urban development topics we have covered in previous chapters have been those which already have some history of effort. In this chapter we shall take a look at some new and future topics.

In this section we will suggest how systems methodology can assist the librarian in making his library serve its users effectively within the constraints of budget and space (*4*). The importance of a library as the storage and dissemination point of societal information cannot be overemphasized. The combined problems of space and budget restrictions and the impeding changes relating to computer applications have made this a difficult time for libraries. These conditions highlight the need to analyze library operations quantitatively and to develop models to represent and predict their behavior. At this point not enough work has been done to assist the librarian adequately in reaching broad policy decisions.

Data acquisition is a problem. Out-of-date books are kept, but old circulation cards are thrown away. Dozens of questions regarding the flow of people through the library and what they do while they are there remain unanswered.

The primary data available to the analyst is the book circulation rate, the number of times a particular book has been borrowed during the year. What must be determined is the demand rate, for if demand is substantially larger than circulation, the users are not being served effectively. Two other important tasks to be done are to predict future circulation from past, and to estimate in-library use of a book from its recorded circulation.

Some things that are relatively easy to measure are the number of users per day, the number of books withdrawn per day, the number of books left on tables per day, the number of books on loan or being repaired at any time, the mean number of books borrowed per user per visit, the use relationship between different book classes, and the mean length of time books stay out on loan. Average values are used because in the management of thousands of books, the objective is to do, on the average, as well as you can.

Because a library may be thought of as having a book inventory, a queuing model can be used to analyze its behavior. The model would encode the relationship between the demand rate for a given book, its circulation rate, and its return rate. A part of the model relating three variables is

$$R = \frac{\mu\lambda}{\mu + \lambda},$$

where
  R = expected value of circulation rate,
  μ = expected value of return rate,
  λ = expected value of demand rate.

Another possible model would be one that determines the demand for a book overtime as a means of detecting the attenuation of popularity of a book. On the average, a book's circulation decays exponentially; however, there are enough exceptions that it is useful to use a Markov process to model the others. First attempts at this model indicate that the model can represent most of the behavior of interest of the circulation history of the book. The Markov model can predict the circulation during the next 5 years of those books with a known circulation this year, and can predict the increase in satisfied users if a duplicate of the book is purchased now.

A search model or a "browsing" model is important because it affects the demand and circulation rate. The probability g that the book of interest is seen is inversely proportional to the number of books N in the section searched. A constant α accounts for the illumination of the shelves and the degree of attentiveness of the browser. During a search of T minutes, when ω is the mean glimpse rate, the chance that the user will not find the book of interest is

$$(1 - g)^{\omega T}$$

and the chance of success is

$$1 - (1 - g)^{\omega T}.$$

Combining and simplifying, the probability P_s of finding the book of interest among N other books during search time T is

$$P_s = 1 - \epsilon^{-\alpha \omega T / N}$$

Typical experimental values for the browsing, rate, $\alpha \omega$, range from 100 to 200 books per minute.

All library classification schemes are attempts to group books that might be sought during one visit, so that the browser can narrow his search area. An interesting possibility is to develop a mathematical programming model in order to arrange it so that books with a frequent commonality of interest to searchers are close together. Until this is done the existing classification systems arrange sections of more or less uniform interest potential for the average browser.

It has been found that an open-shelf subject which contains more than about 5000 volumes becomes too large to be efficiently browsed. Of the several options available to deal with this difficulty, a very direct one is to weed out, or retire, volumes. A variety of models have been explored based on different strategies for reducing the size of oversized sections. Books can be retired based on low circulation, random choice, and age, among others.

The effort in library models should be to devise models that are easy to manipulate and require the least manpower to gather the data. With the models we have discussed, budget allocations could be justified, alternative use of space planned, and new library services instituted.

Urban Blood Banks

A small and often forgotten, but incredibly important, medical resource of any community is its system of blood banks. Blood is collected at one location and point in time, processed, stored, and made available for transfusion to patients at another time and place. Many health problems, including anemia, surgery, burns, and leukemia, require human blood.

Blood banks are an excellent example of one of the many facets of the urban setting which are sufficiently complicated to profit from systems analysis but frequently overlooked in the face of larger issues. Most blood-bank systems have developed without the aid of central coordination and have problems relating to ineffective and inefficient modes of operation (5).

Typically, blood banks are organized into poorly defined regional systems located in a geographically or politically defined area such as a city or a state. The three most common blood-bank problems are

1. High operating costs.
2. A continuing short supply of blood even though nearly a quarter of the stock is lost through outdating.

3. An inability to deal with unpredicted large demands at one or more hospitals.

If a blood-bank system were operating properly there would be minimum delay in responding to requests for blood, the quality of the blood supplied would be high, and the burden of transfusion therapy would be minimized. All these requirements suggest a careful control of blood inventory; however, this is difficult for the following reasons.

1. The supply and demand for blood requires probabilistic descriptions.
2. Half of the blood requested for patients is returned unused.
3. The legal lifetime of blood is 21 days in most places.
4. In order to control inventories, all the interacting banks in the region must be coordinated.

Jennings (5) has investigated the potential for solving these problems.

The aspects of the issues just reviewed which are within the purview of systems analysis are to minimize the frequency of shortages, minimize outdating, and minimize inventory operating costs.

In Figure 11-1 the several sources and uses of blood are illustrated for a single hospital.

Figure 11-1 represents a generalized hospital blood bank, and no two operations are quite the same. Specific information, such as the policy for ordering blood from the central bank, the age of blood received, and the rate at which blood is requested by physicians, is needed to make the model appropriate for a particular facility. The model in Figure 11-1 is too complex for analysis

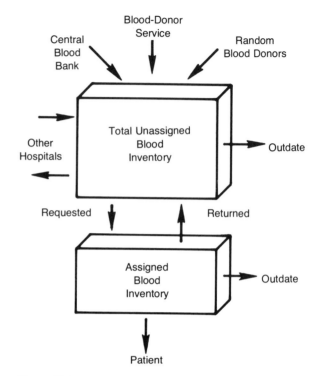

Figure 11-1

based on causal relationships, but a simulation on a digital computer has been done.

The generalized hospital blood bank we have been considering thus far has very little routine operational interactions with other hospital blood banks. The individual blood bank has many components and complex interactions between them; however, little or no interdependence exists with other blood banks. For the regional-blood-banking system model the tradeoff between generality and complexity must be made again. Study has been concentrated on a model of a homogeneous system composed of identical hospital blood banks all operating under the same policies.

Many inventory policies have been studied. An example of a threshold transfer inventory policy is as follows:

1. When the inventory level for a particular blood type falls below a lower threshold of unit, blood is transferred.
2. The lending policy of a blood bank is based on being willing to lend only that blood in excess of a retention level of one unit.
3. When more than one blood lender bank is available, the bank most recently known to have the largest inventory is chosen.
4. When blood is requested, five units are transferred as long as the retention level is not violated.
5. The blood transferred is chosen based on age, the oldest first.

This and many other inventory policies need to be studied using the models. Although the work is still in early stages, some of the results achieved are:

1. The relationship between shortage and outdating performance for systems of various sizes.
2. The nature of support-system capabilities that must be provided.
3. The fact that five cooperating blood banks can achieve most of the gains available in larger, coordinated systems.

Post Office Operations

There have been surprisingly few articles or books written on the subject of postal systems analysis in the professional literature of systems analysis. The post office, like the library and blood bank, is one of these ancillary urban activities which is so integrated into the fabric of city life that it is taken for granted, that is, until there is some difficulty, however minor.

Many post office problems can be traced to population growth and to a former management system that was political in nature. The new U.S. Postal Service is beginning to study fundamental questions of goals, what policies will lead to these goals, how alternative policies will be evaluated, and what models should be developed to assist in this study.

There have been some beginning efforts at developing models of components of the postal system (6). About 32 percent of current postal budgets are used for mail processing, and it is in this area that we want to consider model evolution.

Mail processing refers to all types of mail; however, for discussion purposes we will restrict ourselves to letter-mail processing. The three types of letter mail which define the three major activities of mail processing are

1. Collection and acceptance mail.
2. Transit mail.
3. Incoming mail.

Collection and acceptance mail is collected at mailboxes or brought directly to the post office. Transit mail is processed by the local post office but is originated in and destined for other post offices. Incoming mail is a part of local delivery, but it originated in another city.

All the letter-sorting operations used to prepare these three types of mail for distribution are manual except the one done by the letter-sorting machine (LSM). At each console of the LSM, the operator sends each letter to the appropriate bin, based on reading the address.

Two letter-mail-processing problems have been identified. The first problem is acquisition of the data necessary to efficiently distribute the total volumes of the three mail types among the five sorting operations. The second problem is how to evaluate savings from the use of mechanized sorting equipment in dealing with large volume variations and performance requirements.

Technical improvements in mail sorting have been very slow to come. Only a few optical character readers exist at this time. These electronic devices are able to scan a few types of envelopes and automatically direct the envelopes to the proper output bin.

One aspect of the difficulty of letter-mail processing is the fact that 40 to 60 percent of the collection and acceptance mail is received during the 4-hour period from 4 P.M. to 8 P.M. This severe peaking effect requires a mechanized sorting capacity which will be underutilized much of the remaining time. This highly variable demand for letter processing has caused post offices to cling to the practice of hiring low-cost hourly manual labor to deal with the high demand workloads.

Markov Postal Model

The first model we will discuss takes into account the variable demand problem when estimating the cost savings of equipment. A simple Markov model will be used for determining the sorting costs per letter as a function of mail type and initial operation. The model assumptions are:

1. The probability that a letter will go to the next operation depends only on the previous operation (the Markov assumption).
2. The average value of sorting costs are meaningful; that is, short-term variations cancel out over the long run.
3. All errors result from missorts in the LSM operation.

Using these assumptions the mathematical model for determining the total sorting cost as a function of mail type and initial operation is

$$C_i(k) = \sum_{j=1}^{5} n_{ij}(k)(L_j/R_j),$$

where

L_j = average labor cost per man-hour for operation i_j, dollars/man-hour,

R_j = average productivity of operation i_j, pieces/man-hour,

$n_{ij}(k)$ = average number of times a type k letter is handled in operation j, given that it was first handled in operation i,

$C_i(k)$ = average sorting cost for type k mail that is initially handled by operation i_j, dollars/piece.

This model has been used to find the sorting costs in post offices. Some of the interesting results that have been observed are: corresponding cost elements vary significantly between post offices, costs vary widely by mail type, and costs for a particular mail type in a particular office vary considerably by initial operation. The possibility of developing a postal cost-accounting system using this Markov model and a new data-collection system are being considered.

Linear Programming Postal Model

The intention of the linear programming postal model is to be able to estimate realistic upper bounds on the cost savings derivable from sorting machinery. Current service requirements and projected demand requirements are considered in order to predict estimates of the minimum total sorting costs for a combined manual and mechanized system. The assumptions for this model are:

1. The cost per piece as a function of mail type and initial operation is known.
2. The three mail types are collection and acceptance, transit, and incoming. The arrival rates are known for all three mail types. Straight-line approximations of the three arrival rates are given in Figure 11-2. All curves are cumulative.

The sorting costs of the combination manual and mechanized system are minimized by the linear programming model. The objective function of the model is

$$Z(V) = \min \sum_{i=1}^{3} \sum_{j=1}^{24} \left[C_p(i) V_p(i,j) + C_m(i) V_m(i,j) \right],$$

where

$Z(V)$ = minimum total cost of sorting a daily volume V,

$C_p(i)$ = cost of manual sorting for type i mail,

$C_m(i)$ = cost of machine sorting for type i mail,

$V_p(i, j)$ = volume of type i mail processed during hour j by the manual system,

$V_m(i, j)$ = volume of type i mail processed during hour j by the mechanized system.

The objective function, then, is merely the summation of the hourly costs of manual and mechanized sorting for each of the three mail types.

A number of constraints must be imposed on on the objective function. The first constraint is due to the capacity of the mechanized system:

$$\sum_{i=1}^{3} \frac{V_m(i, j)}{R_m(i)} \leq 1 \qquad \text{for all } j,$$

where

$R_m(i)$ = the number of pieces of mail per hour which the mechanized system is capable of processing.

Any solution to the linear program must ensure that the mail processed since the last processing deadline must not exceed the cumulative arrival rate. The form of this constraint is

$$\sum_{t_i=1}^{k} \left[V_m(i, t_i) + V_p(i, t_i) \right] \leq \sum_{t_i=1}^{k} A(i, t_i) V,$$

where

V = total daily volume of mail,

t_i = hours since the last processing deadline for type i mail,

$A(i, t_i)$ = fraction of total daily mail volume which is type i mail arriving during hour t_i.

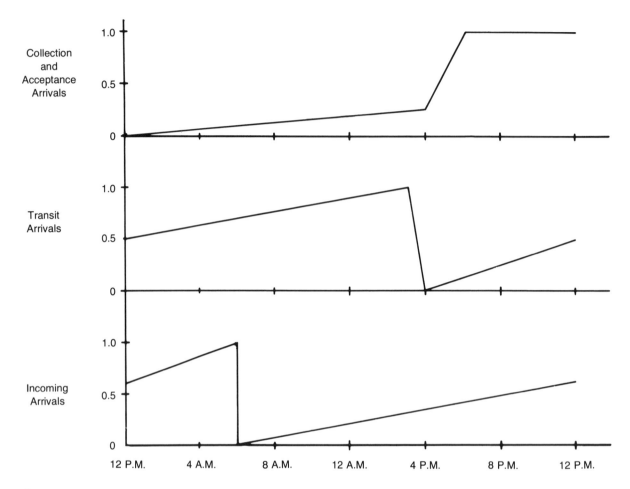

Figure 11-2

If $Z(V)$ is summed over the year, the annual sorting cost is obtained.

The last aspect of the model was an attempt to predict the minimum sorting cost for future years, including the effects of increasing mail volume, increasing postal employees' wages, and the time value of money. The minimum sorting cost for a future year, Y_n, is

$$Y_n = \frac{(1 + \omega)^{n-1}}{(1 + r)^{n-1}} \sum_{T=1}^{365} Z\left[(1 + v)^{n-1}V_T\right],$$

where

ω = postal-wage-increase factor,
r = annual discount rate,
v = annual mail-volume-increase factor,
V_T = volume of mail on the corresponding day of the current year.

The model can be adopted for mechanized systems with different initial costs, different capacities, and variations in performance. For this reason the model can be used to determine the optimal size of a mechanized system.

In addition to assisting in the minimization of sorting costs, the model can be used for improving the scheduling efficiency of existing mechanized equipment, and making investment decisions for new mechanized sorting equipment.

Industrial Dynamics

It is our judgment that the process and strategy employed in producing a model is the most critical aspect of the systems approach. The model, in effect, looks back to the precise verbal description from which it was formed; it looks ahead to be sure that suitable analytic tools are available, and the model looks even further ahead to be sure that any results obtained bear directly on the problem that is being solved.

Undoubtedly the best-known modeler in the country, and the one whose work has had the greatest impact, is Jay W. Forrester of the Sloan School of Management at the Massachusetts In-

stitute of Technology. It is appropriate that in a chapter devoted to the future of the systems approach, Forrester's work appears. His published works constitute the most impressive collection devoted to systems modeling. In this first section discussing Forrester's work, the basic concepts of his modeling strategy are discussed as applied to industrial problems (7).

Forrester defines his methodology in the following way (8):

Industrial Dynamics is the study of the information feedback characteristics of industrial activity to show how organizational structure, amplification (in policies) and time delays (in decisions and interactions) interact to influence the success of the enterprise. It treats the interaction between flows of information, money, orders, materials, personnel, and capital equipment in a company, an industry, or a national economy.

His approach is to form a simulation based upon a mathematical replication of the actual firm. The following is the procedure for constructing and running his type of simulation:

1. Identify a problem.
2. Determine the factors that appear to interact to create the observed symptoms.
3. Trace the cause-and-effect information feedback loops that link decisions to action to resulting information changes and to new decisions.
4. Formulate acceptable decision policies that describe how decisions result from the available information.
5. Model the decision policies, information sources, and interactions of the system components.

6. Use the model to generate system behavior.

7. Compare model results against historically available knowledge about the system.

8. Revise the model until a better match is realized to existing data.

9. Modify the model by changing organizational relationships and policies which can be altered in the actual system to identify the changes that improve system behavior.

10. Implement the results from the model to make changes in the real system which will improve performance.

Here again is another expanded and slightly modified version of the systems approach. The simulation utilized by Forrester is completely quantitative based on the conviction that managers in the real world apply rather simple decision rules which lend themselves to quantification. The model is formed based on empirical performance data and interviews with managers regarding the rules they apply in making decisions. The model is developed by concentrating on the factors that determine the characteristics of information feedback, systems structure, amplification, and delays. Flowcharts are utilized to interpret verbal descriptions and observations. Levels and flows of information, money, orders, materials, personnel, and capital equipment are portrayed on the flow chart. An example of a level equation is

new level = previous level + (rate of inflow − rate of outflow) × time increment.

Another type of equation, the rate equation, serves two purposes. The formation of rate equations is equivalent to basing decisions on only the information available about past performance of the system. Rate equations serve the purpose of determining system dynamics, such as the delay in transit of a particular flow between two levels. The form of a delay rate equation is

$$\text{outflow rate during a specified time} = \frac{\text{amount stored in the delay at the beginning of that time}}{\text{average length of time to traverse the delay}}$$

This type of delay equation produces an exponentially decaying rate of outflow in response to a sudden input.

These equations and others form the industrial dynamics model, and a special compiler called DYNAMO was developed at MIT for solutions of problems using this model. If the data and the model are given in the appropriate form, the DYNAMO compiler checks the equations for logical consistency, solves the system of equations, tabulates data, and graphically plots the variables specified. The special-purpose compiler is extremely convenient and reduces the amount of effort required in a simulation run. The model is validated by observing whether the model reproduces or predicts such systems characteristics as stability, oscillation, growth, average period between peaks, general time relationships between changing variables, and tendency to attenuate externally imposed disturbances. Therefore, the model is validated by dynamic characteristics of the system rather than a correspondence between predicted and observed time-phased behavior.

This simulation approach called industrial dynamics has been met with various reactions. Although many researchers resist the highly quantitative aspects, all agree that there is a significant benefit from the process of flowchart modeling of the firm. These preliminaries to the simulation have an almost therapeutic effect on the managers, forcing them to crystallize decision-making processes and to order their thoughts according to a systematic information-feedback model. All investigators agree in the benefits of this step, and, of course, the costs of flowcharting are much less than an actual simulation. Many management problems of the firm have been successfully described through feedback representation, and application to other problems will depend on the particular circumstances.

Forrester identifies industrial dynamics as a philosophy of structure in systems (9). He believes further that it is becoming a body of principles that relates structure to behavior.

Forrester-Model Formalism

Forrester's work represents a direct attempt to model the dynamics of social and economic processes. For this reason it is potentially so important and is in contrast to the statistical time-series regression-analysis approach which yields little insight into the actual structure of the system. It is possible, however, that incorrect conclusions can be made from Forrester-type models and, in fact, in the limit such models may be adjusted to yield nearly any result (10).

A possible form of the general model equations is as n first-order nonlinear ordinary differential equations:

$$\frac{d}{dt}y_j(t) = y_j(t)(K_{j+}a_{j+} - K_{j-}a_{j-}), \qquad j = 1, ..., n,$$

where

y_j = output or level variables which are important in the system,

K_{j+} and K_{j-} = nominal percentage rate of increase or decrease in variable y_j,

a_{j+} and a_{j-} = "incentives" which modify the nominal influx or efflux rates according to existing conditions in the system.

The incentives can be written as the product of a series of multipliers, each of which quantifies the effect of a single particular factor:

$$a_{j+} = \prod_{k=1}^{l} M_{jk+}(y, l),$$

$$a_{j-} = \prod_{k=1}^{l} M_{jk-}(y, l),$$

where

$M_{jk+(-)}$ = multipliers that are multidimensional functions of exogonous variables l as well as the system-level variables y.

The multiplier functions just defined are a major contribution of Forrester's formalism. When these multipliers are combined in a multiplicative fashion as shown above, they reflect conditions of the state of the system relative to conditions identical to the nominal conditions, while a value greater than 1 would reflect an increased tendency to increase the y_k. The multiplier functions can be represented as piecewise linear curves or as certain elementary analytical functions, such as a bell-shaped curve.

As a means of emphasizing the Forrester model and especially the multiplier functions we will look briefly at the Tahoe GUESS model. The Lake Tahoe basin, a recreational area on the California–Nevada state line near Reno, Nevada, has been modeled using the Forrester formalism. The model, entitled Tahoe GUESS (for Geo-Urban-Eco-System-Simulation), represents the Tahoe basin as an emerging urban area which does not markedly influence the growth of surrounding urban centers. A major difficulty in socioeconomic modeling is the quantification of subsystem interactions and relations among variables for systems. A further complication is the fact that the people most knowledgeable about social systems, that is, the social scientists and political decision makers, have little background in mathematical quantification. In Tahoe GUESS a research sociologist as a member of the analysis team helped select the level variables for the model. They are

y_1 = permanent residents,
y_2 = part-time residents,
y_3 = residences,
y_4 = hotel and motel accomodations,
y_5 = camping facilities,
y_6 = resident-supported business,
y_7 = tourist-supported business,
y_8 = environmentally related tourists,
y_9 = gaming-related tourists.

The sociologist also described qualitatively, based on his research and intuition, his hypotheses of the system interactions. Next the hypotheses were quantified graphically as multiplier functions. Twenty-five multipliers were used to relate the nine level variables. As an example, the availability of housing is a critical factor influencing the incentive for potential residents to move into the Tahoe basin. Based on expert opinion the nature of this influence has been hypothesized to be of the form given in Figure 11-3.

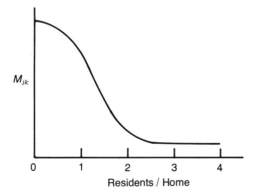

M_{jk}

Residents / Home

Figure 11-3

In the GUESS model M_{jk} from Figure 11-3 is designated M_{13}, that is, the third multiplier affecting the first-level variable. The multiplier curve clearly portrays the decrease in incentive to move when the ratio of residents to homes is large. It is important to emphasize that this curve in Figure 11-3 is sketched based on interrogating a social scientist with a deep understanding of the research setting. The curve which resulted can be represented closely by a function of the form

$$M_{13} = A \arctan \frac{B(x/xn)}{\left[1 - (x/xn)\right]^2} + C_m,$$

where

x = ratio of residents to homes,

xn = value of x to which M_{13} is most sensitive (intuitively near $x = 1.0$),

A = difference between the maximum and minimum values of M_{13},

C_m = minimum value of M_{13}.

All curve parameters are adjusted in order to quantify the hypothesis in an intuitive fashion.

Predictions of the values of variables of GUESS were obtained for a 20-year period. The simulated time histories suggested many aspects of the rapid urbanization of the Tahoe basin and are part of the information now available to the appropriate authorities.

The GUESS model was developed in reference (*10*) using a systematic procedure which is a useful implementation of the Forrester formalism. The steps of the procedure are as follows:

1. With the assistance of experts, choose the important system-level variables.
2. In collaboration with experts, qualitatively formulate the model hypotheses, then quantify them as multiplier functions.
3. Collect all available data on level variables and rates of change of level variables.
4. Define nominal conditions.
5. Adjust the model to account for the trend in the data.
6. Determine those parameters to which the model is most sensitive and which can be least accurately estimated from the data.

7. Make an initial estimate of the change in the level variables, and then use an identification scheme to determine the parameters of step (*6*).

Urban Dynamics

Jay Forrester's second major contribution to the modeling literature was his book *Urban Dynamics* (*11*), in which he constructs a simulation model of the housing, industry, and population of a single urban area. This model is used to draw conclusions about the relative effectiveness of various public policies. Population is divided into three groups: management and professional workers, skilled workers, and unskilled workers. Housing is divided into categories suitable for the three groups. Industry provides employment and competes with housing for the land area. The tax structure is limited, including only real estate taxes against housing and industry. For the urban model, the equations describe the construction, decline, and demolition of industry and housing, and changes in the population levels for the various groups. For example, the rate of change of the number of unskilled workers, U, is (*12*)

$$U(\text{this year}) - U(\text{last year}) = UA - UD + LTUN + UB,$$

where

UA = rate of arrival of unskilled workers into the city,

UD = rate of departure of unskilled workers from the city,

$LTUN$ = net flow via social mobility between the skilled and unskilled groups,

UB = net rate of births minus deaths.

The flow rates that describe the arrivals and departures of workers are given by

$$UD = \text{constant} \cdot U \div ATT.$$

This expression encodes the information that the number of unskilled workers departing each year is proportional to the number of people in the unskilled group divided by a number, ATT, which represents the attractiveness of the city for the U group.

In the same way,

$$UA = (\text{another constant}) \cdot (U + L) \cdot ATTP.$$

In this equation the arrival rate is proportional to the total number of workers and to the attractiveness of the city as perceived by outsiders. To give a precise numerical meaning to "attractiveness," an approach similar to the one used in "incentive" of the previous section is utilized. The magnitude of the various forces which draw people to the city is "guessed." Mathematically, attractiveness is a product of several factors describing each of several components. The magnitude of the quantity, attractiveness, for unskilled workers grows

1. As their economic opportunity grows.
2. As the density in their housing diminishes.
3. As their unemployment rate diminishes.
4. As the public expenditure per capita increases.

5. As the underemployed housing program produces superior housing units.

In Figure 11-4 appears Forrester's estimate of the dependence of the attractiveness upon residential density in the housing for the underemployed. As mentioned before, these perceived magnitudes of experts are a critical ingredient in the Forrester approach. Forrester separates perceived from actual attractiveness. The perceived attractiveness which governs immigration of unskilled workers is assumed to differ from the actual attractiveness because outsiders cannot keep abreast of changes in the city. As before, rates of change in the model are defined, and these rate equations are solved to find the development of the city. The model allows the time development of the filling up of the city until an equilibrium point is reached. Some of the success of the model output through this process is indicated by the city defects which appear. For example, one of the urban models shows a 45 percent unemployment rate among the unskilled, a

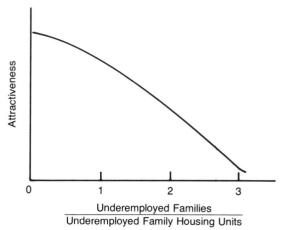

Figure 11-4

lack of skilled labor, and an excessively high real estate tax, required to pay for the needs of the unskilled workers and their families.

With a Forrester urban model of the type we have been discussing, alternative public policies can be tested to ease the urban problems that have been identified. Examples of two alternative policies which Forrester discusses in *Urban Dynamics* are the job-training program and the clearance program. The purpose of the job-training program is to train 5 percent of the unskilled population each year for skilled jobs. This policy is encoded in the urban model by putting an extra 5 percent of the unskilled group (U) into the skilled group (L) in each year's interval. For the clearance program an extra 5 percent of the unskilled group's housing is demolished each year and resistance to the construction of housing for the skilled group is doubled. The computer program is then run starting with the city at equilibrium, to see what changes are produced.

By inference, Forrester's readers have concluded that his goals for the city are:

1. Reduce the proportion of unskilled workers in the population.
2. Reduce taxes.
3. Increase the net upward social mobility from unskilled to skilled.

If these three goals are examined in the light of the clearance program, the results are satisfying. The following narrative describes the effects of the clearance program. Through the program, housing for the unskilled is demolished while simultaneously the limiting of construction of skilled worker housing discourages the natural transition of such housing to the unskilled group. These two effects cause unskilled workers to leave the city.

With fewer poor people present, the city's taxes can be lowered, and the land freed by the demolition of housing can be utilized for the construction of new industry. Job opportunities will increase as industry increases and net upward mobility will rise.

The equivalent narrative for the training program is as follows. The first effect of the training program is an increase in skilled labor and a decrease in unskilled labor. However, the existence of the training program makes the city more attractive to unskilled workers not living there and they migrate to the city. As a consequence, the training program produces only a small change in the proportion of unskilled workers and in tax rates. For these reasons, when examined against Forrester's urban goals, the clearance program is to be desired over the job-training program. It is important to remember that the previous narratives were constructed based on the computer-model output results for the several variables.

Forrester's attractiveness functions have the impact of a social index. The attractiveness functions give numerical estimates of the worth of the city as perceived by various groups. The changes in attractiveness resulting from proposed programs can be computed from the model. In this way, the numerical worth of any public program to the involved groups can be estimated. In Figure 11-5, the changes in urban attractiveness to both unskilled and skilled workers due to a job-training program are sketched from the computer-model output. Note that the attractiveness measure is normalized so that each attractiveness number is equal to one in the steady-state solution.

A broader view of the Forrester urban model considers not only the urban area of interest but

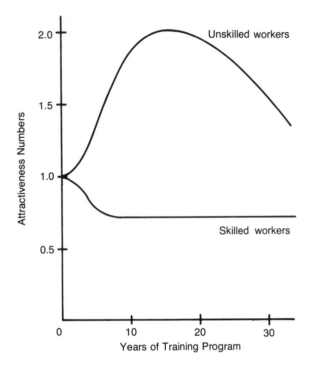

Figure 11-5

also the fact that public programs have important effects upon the rest of the nation because they change migration patterns between cities. It seems reasonable that a job-training program would attract unskilled workers from other cities and that a clearance program would force workers into other areas. So, the "solution" to the urban problem in one area may influence the intensity of urban problems in nearby areas.

Modeling an Urban Area

There is little controversy regarding the fact that Forrester's work is a pioneering effort. How-ever, controversy has arisen regarding other aspects of the model. One concern is directed toward the complexity of the urban model, and Stonebraker has done interesting work in simplifying Forrester's model of an urban area (*13*). Stonebraker capitalized on the fact that the urban model was very stable and insensitive to parameter changes. These qualities suggest that the original Forrester model might be simplified from a set of model features which are burdensome to consider as a whole. In Figure 11-6 a diagram of Forrester's urban model is given which has nine important level variables. The three vertical levels in the figure correspond to business, housing, and people, respectively, top to bottom. The flow paths between the states are indicated by arrows and the variables moving along these flow paths are identified by code letters. A key to the code letters in Figure 11-6 is given next.

J	Jobs	G	Growth rate
H	Housing	A	Adequacy of current housing
T	Taxes	UM	Upward mobility
D	Delay	S	Social compensation
L	Land	W	Labor available to work
B	Births–deaths	C	Labor available for construction
M	Managers available	HP	Low-cost housing program

The simplification of Figure 11-6 given in Figure 11-7 was accomplished by eliminating variables in three ways.

In the simplified Figure 11-7, the following set of code letters is used.

J	Jobs	M	Managers available
H	Housing	W	Labor available to work
L	Land	A	Adequacy of current housing

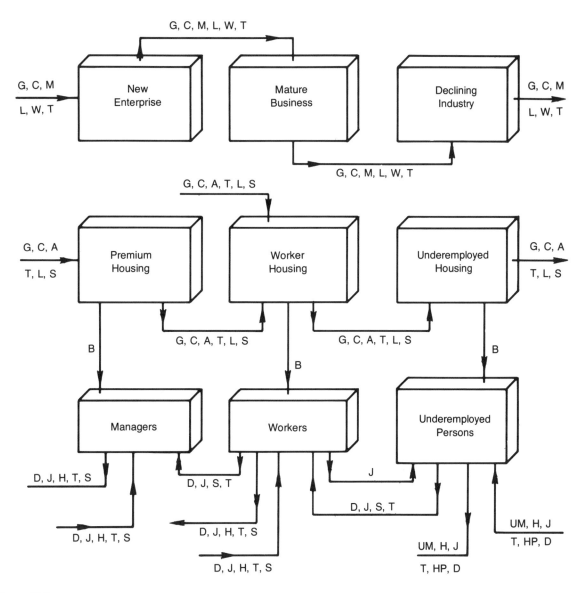

Figure 11-6

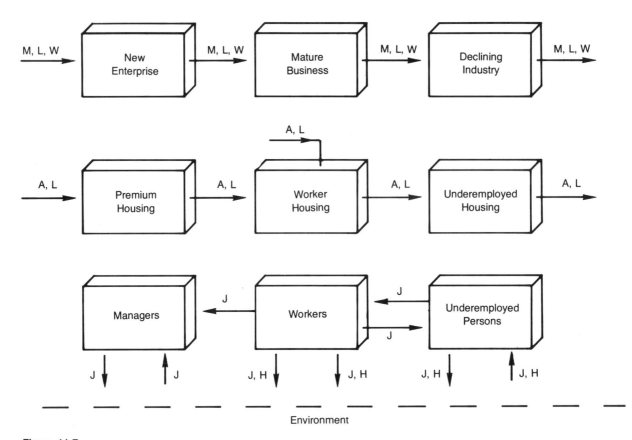

Figure 11-7

The simplification from Figure 11-6 to Figure 11-7 was accomplished using three approaches:

1. The model was studied to determine which variables were unimportant to the behavior of the model.
2. Groups of interacting variables were removed by trial and error.
3. Sensitivity studies were used to identify unimportant variables.

In Figure 11-6 the arrival of underemployed people from the external environment depends on the availability of jobs, the availability of housing, the tax rate, the existence of a low-cost housing program, the observed upward mobility of underemployed persons in the city, all of these conditions being perceived after a delay for news to reach the environment and for people there to act on it. In his simplified model, Stonebraker allows the labor classes to move to and from the environment solely as a function of housing and jobs. The point of Stonebraker's effort has been to attempt to deal with the complexity of the Forrester urban model; in doing so, he has provided us with a unique closer look at the model.

It is interesting to note that while Forrester's model is being simplified in one setting, other critics are proposing the addition of time-varying technology, elimination of the fixed land area, age-specific migration rates, and other factors to the model.

Counterintuitive Behavior of Social Systems

In this chapter devoted to the future, we are spending a great deal of time discussing Forrester's work. A natural part of this is an exposition regarding social systems taken from Forrester's own writings (*14*). In a broader context, there is concern over the quality of urban life, national policy for urban growth, and the unexpected, ineffective, or detrimental results often generated by government programs in these areas.

The systems approach, specifically Forrester's work, is aimed at identifying the orderly processes that create our social system, and understanding the human judgment and intuition that frequently lead people to wrong decisions when faced with complex and highly interacting systems. The basic theme of the modeler is that we can now construct, in the laboratory, realistic models of social systems so that new laws and programs can be tried out on them in advance. The objection heard most often is that our knowl-

edge of social systems is insufficient to do it; however, in a sea of controversy and criticism, Forrester is doing it.

Under the label of "systems dynamics," researchers have studied many corporations which were having severe and well-known difficulties. Discussions with the key decision makers revealed that they perceived their environment correctly; they knew what they were trying to accomplish; they knew the crises that would force certain actions; and they were sensitive to the power structure of the organization, to traditions, and to their own personal goals and welfare. The company problem solving was being done in good conscience, and particular policies were being followed on the presumption that they would alleviate the difficulties. The analyst has used the policy information with a computer model to show that often, in concert, the extant company policies actually cause trouble. Internal difficulties are created independent of external problems, and in combination a downward spiral develops in which the presumed solution makes the difficulty worse and causes an increased effort for the presumed solution.

Because of the similarity between the corporate and the urban situations, four common urban problem-solving programs were examined using the Forrester approach.

1. Jobs are created by bussing the unemployed to the suburbs or by providing governmental jobs.
2. A training program is used to increase the skills of the lowest income group.
3. A federal subsidy is given as financial aid to the depressed city.
4. Low-cost housing is constructed.

Forrester makes the following comment regarding these four programs:

All of these (programs) are shown to be between neutral and detrimental almost irrespective of the criteria used for judgment. They range from ineffective to harmful judged either by their effect on the economic health of the city or by their long range effect on the low income population of the city.

It is Forrester's position that it is *excess* housing in the low-income category rather than the commonly presumed housing shortage which is the primary cause of depressed areas in the cities. A narrative that supports this position is as follows: Old buildings persist because of the effect of legal and tax structures, but as industrial buildings decay, their employment opportunities drop. When residential buildings age, their occupancy by lower-income groups is at a higher population density. Combined, we have declining jobs and increasing population while the buildings age. With more housing available to low-income groups, they continue coming to the city until their population exceeds the opportunities, and then the standard of living declines to stop the flow. With a low income to the area, excess housing falls into disrepair and is abandoned. Therefore, coexisting are overcrowded buildings and buildings that are empty because the economy of the area cannot support all the residential structures. These abandoned buildings occupy potential industrial land and offer housing if economic factors change. Both factors appear undesirable. The social system just described regulates itself by responding to a change that would raise the standard of living by removing the economic pressure only enough to permit a population increase, which again lowers the standard of living to the barely tolerable level.

This narrative represents a *Forreser result*, an excellent example of how a simulation model can lead to theory development. The idea of the narrative could be extended to include the near-equilibrium that affects population in different areas of the country. Migration forces can be analyzed using this approach.

Forrester's writings contain many primitive theories of urban processes. Often these theories are controversial. For example, consider the following statement by Forrester (*14*):

Over the last several decades the country has slipped into a set of attitudes about our cities that are leading to actions that have become an integral part of the system that is generating greater troubles. If we were malicious and wanted to create urban slums, trap low-income people in ghetto areas, and increase the number of people on welfare, we could do little better than follow the present policies. The trend toward stressing income and sales taxes and away from the real estate tax encourages old buildings to remain in place and block self-renewal. The concessions in the income tax laws to encourage low-income housing will in the long run actually increase the total low-income population of the country. The highway expenditures and the government loans for suburban housing have made it easier for higher-income groups to abandon urban areas than to revive them. The pressure to expand the areas incorporated by urban government, in an effort to expand the revenue base, have been more than offset by lowered administrative efficiency, more citizen frustration, and the accelerated decline that is triggered in the annexed areas. The belief that more money will solve urban problems has taken attention away from correcting the underlying causes and has instead allowed the problems to grow to the limit of the available money, whatever that amount might be.

It takes little imagination to realize how many professionals and institutions are challanged by this kind of statement. And Forrester's model-building, quantitative data-oriented approach cannot be ignored. Sometimes, entire issues of professional journals, such as reference (15) are devoted to discussions of Forrester's work.

Three reasons for the counterintuitive behavior of social systems are:

1. Social systems are essentially insensitive to most policy changes that people choose in an effort to alter the behavior of the system. At times it appears that attention is drawn to some point in an urban system at which attempts to intervene are doomed to failure. The case we discussed earlier is an example. In an attempt to relieve the suffering of a population depressed because of inadequate housing, more housing is built and an increase of population traps more people in a depressed social system.

2. There are a few sensitive influence points through which the behavior of the system can be changed. Their location is usually unexpected, and a further complication is that even when identified, a person guided by intuition and judgment is likely to alter the system in the wrong direction. For example, if Forrester's conjecture is correct, it appears that low-income housing must be reduced rather than increased in order to revive the economy of a city and make it a better place to live.

3. There is usually a fundamental conflict between the short-term and long-term consequences of a policy change. Short-run improvements usually degrade the system in the long run. And a program designed to produce long-run improvements often depresses the system at the outset.

City Politics

As we near the end of this final chapter, in retrospect there have been three urban factors which have been difficult to handle: economical, social, and political. We have frequently touched economical and social factors in our discussions, but almost never political factors. This is astonishing when you consider that in spite of literally endless information on a system, the entire future of the system can be changed by one politically motivated decision by a decision maker. Little effort has been directed to simulating the political subsystem of a city. In this section we will review a political model that assigns utility functions and decision rules to the aldermen and mayor of the city, who take into account both potential votes and jobs and favor patronage (16). The model is in preliminary form and indicates how the political system might operate if decisions on all aspects of political behavior had to be made consciously.

The description of political processes is based on case studies of urban politics taken from the literature. The model describes a city of 1 million voters, divided into 20 wards, each with a total vote of 50,000 per election. Aldermen are chosen from each ward on a partisan basis, and bargaining among the aldermen occurs in order to make decisions. A small group of aldermen and a strong Democratic machine are relatively influential in running the City Council. A partisan city-wide election selects the Democratic mayor, who is the leader of the political machine both in the

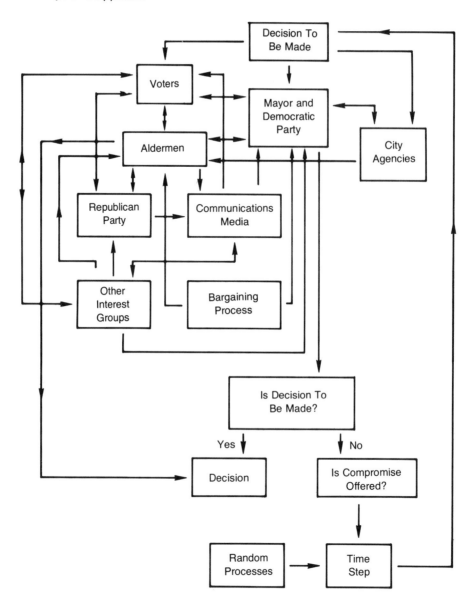

Figure 11-8

city and the state. Many special-interest groups are interested in politics, including business and industrial interests, unions, newspapers, churches, civic organizations, organized crime, and other groups which emerge under the pressures of specific controversies. The model conditions are developed from Chicago and a few similar political machine cities, but flexibility allows a wide variety of cities to be accomodated.

Figure 11-8 shows the overall structure of the political model, with emphasis on the mayor and the aldermen. The aldermen chose their position on issues on the basis of votes gained or lost and on the number of additional positions of favors he can gain for his constituents and allies. The input to the political model in Figure 11-8 is the decision to be made and is assumed to come from outside the model. The mayor determines his own position on an issue, then decides when and whether to intervene according to a predetermined set of decision rules. He can force a decision or delay it, bargain or impede the bargaining process; and initiate compromises on issues.

The communications media exerts pressure on the mayor and aldermen and are used by them to publicize their position and to keep up voter interest in the issue.

When the model is operated, it progresses in time until the proposition is decided. Clearly, with a model this complicated, the choice of parameters and functional relationships plays a crucial role in the behavior of the model. Both data and description were used to gain insight into each process of Figure 11-8. Each process was studied to identify key variables, to detect directions of and bounds upon the effects, to determine an indication of the general form of the relationship, and to develop a feeling for the statistical description of the intermediate and the ultimate results of the process.

As an example let us look at the attitude-influence sector of the political-simulation model. Communications from the media, both information and opinion, may be received directly or may reach a recipient indirectly through discussions with family, friends, and co-workers. Following this, a period of time elapses for more reception, evaluation, modification, and possibly retransmission before the original message is incorporated into the beliefs and attitudes of each individual. At least three effects follow the communication process: underlying attitudes surface, some ideas are changed, and an awareness is reached that some attitudes are inconsistent with deeply held beliefs. These processes are shown in Figure 11-9.

The description given in Figure 11-9 can be translated into a model in the following way. Information and opinion from a number of basic sources of varying credibility are received by the ward. People weigh this information according to its credibility. The effective level of coverage is modified by the salience of the issue to the people in the ward. The model uses an index E which is a measure of the impact of information upon the welfare of the voters in the ward as they perceive it. E can range from -3, a strong anticipated negative impact on welfare, to $+3$, an equally strong positive impact. A second index, I, is a measure of the extent to which voters in the ward, in choosing their positions as an issue, tend to take into account the welfare of people in other wards. For fairly well educated middle and wealthier classes, I approaches the maximum of

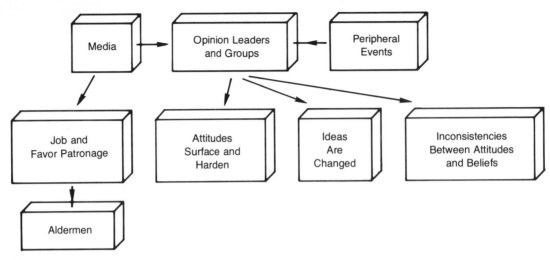

Figure 11-9

2. For an economically depressed poorly educated ward, I will be 0, or even negative. In the attitude model, the index ETRUE represents the underlying value of E. A change in ETRUE depends on the magnitude and salience of the information. The index ITRUE represents the hardening, changing, and inconsistent resolving of beliefs and ideas.

The level of news coverage is determined by assigning weights ranging from 1 to 3 to events. A measure of the level of coverage is the sum of the squares of these weights. The level of coverage is influenced by a number of stochastic elements. For example, there is a 5 percent chance at each step that an alderman with a strong ($I2$) position will do something newsworthy.

The E and I indices of a single ward show the impact of news by taking a weighted average of the level of coverage associated with each medium or interest group. The effective level of opinion is found by weighing each information contribution by relative attention, by the position taken by the group, and by a factor that reduces the weight of the source the more its position differs from that of the natural position of the ward. These paragraphs have given some of the details associated with the description of the way in which media and interest groups influence public opinion. Using this information, the description given in Figure 11-9 is translated into the attitude-sector model in Figure 11-10. Figure 11-10 is one of the complex subsystems of the political model and is presented as an indicator of the complexity of the entire model.

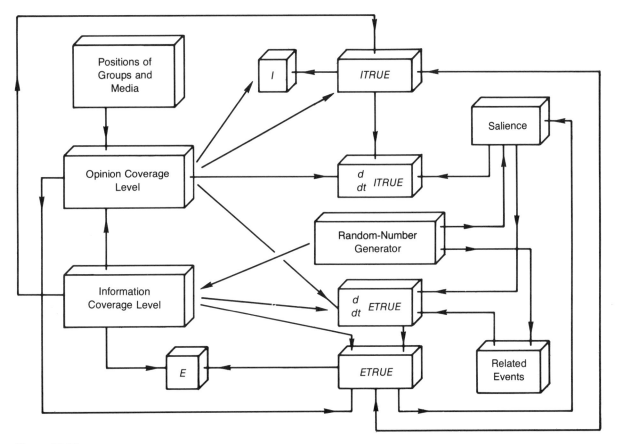

Figure 11-10

At the time of this writing, the political model was not validated, but because the parameters and functions used in the model are based on descriptive and qualitative inputs derived from case studies of urban political processes, the task is hopefully not difficult. It is anticipated that the model, as a whole, will represent the generally accepted views of the political process in an urban environment.

The Frontier

The promise and results of the systems approach have been studied in every chapter. We have written in many ways about taking the separate parts of an urban system and combining them in a model to learn the consequences. The hypotheses upon which the subsystems are based may at first be no more correct than the ones we are using in our intuitive thinking. However, the evolution of the model requires these hypotheses to be stated more explicitly. A hazy mental model is transformed into an unambiguous model *to which all have access.* As available information grows, assumptions can be rapidly improved. With a stated set of assumptions, the model, perhaps represented as a computer program, will trace the resulting consequences without doubt or error.

After earlier periods when human pioneering and creativity have been focused on geographical exploration, formation of governments, literature, science, and technology, it is exciting and pertinent to commit ourselves to the understanding of social processes. The message of this book is that the means to make this commitment are visible.

It appears that if we follow our intuition and predispositions in social decision making, the trends of the past will continue into deepening difficulties. Using the systems approach, we can set up research and educational programs which are now possible but which have not yet been developed. As an outcome of this effort, we can expect a far sounder basis for action.

References

(*1*) D. A. Curry, A Systems Approach to Societal Problems, presented at the Joint National Conference on Major Systems, Anaheim, Calif., Oct. 28, 1971.

(*2*) U.S. Department of Health, Education, and Welfare, Toward a Social Report, Government Printing Office, Washington, D.C., 1969.

(*3*) National Industrial Conference Board and the Opinion Research Corporation, Perspectives for the '70's and '80's, an experimental forecast conducted in 1968, 1970.

(*4*) P. M. Morse, Library Models, Chap. 12 in *Analysis of Public Systems,* A. W. Drake, R. L. Keeney, and P. M. Morse, eds., The MIT Press, Cambridge, Mass., 1972.

(*5*) J. B. Jennings, Blood Bank Inventory Control, Chap. 11 in *Analysis of Public Systems,* A. W. Drake, R. L. Keeney, and P. M. Morse, eds., The MIT Press, Cambridge, Mass., 1972.

(*6*) C. C. McBride, Post Office Mail Processing Operations, in *Analysis of Public Systems,* by A. W. Drake, R. L. Keeney, and P. M. Morse, eds., The MIT Press, Cambridge, Mass., 1972.

(*7*) J. W. Forrester, *Industrial Dynamics,* The MIT Press, Cambridge, Mass., 1961.

(*8*) H. I. Ansoff and D. P. Slevin, An Appreciation of Industrial Dynamics, *Management Sci.,* **14**(7), (1968).

(*9*) J. W. Forrester, Industrial Dynamics—After the First Decade, *Management Sci.,* **14**(7), (1968).

(*10*) J. W. Young, W. F. Arnold, and J. W. Brewer, Parameter Identification and Dynamic Models of Socioeconomic Phenomena, *IEEE Trans. Systems, Man, Cybernetics,* **SMC-2**(4), (1972).

(*11*) J. W. Forrester, *Urban Dynamics,* The MIT Press, Cambridge, Mass., 1969.

(*12*) L. P. Kadanoff, From Simulation Model to Public Policy, *Am. Scientist,* **60**(1), (1972).

(*13*) M. Stonebraker, A Simplification of Forrester's Model of an Urban Area, *IEEE Trans. Systems, Man, Cybernetics,* **SMC-2**(4), (1972).

(*14*) J. W. Forrester, Counterintuitive Behavior of Social Systems, *Technol. Rev.,* **73**(3), (1971).

(*15*) *IEEE Trans. Systems, Man, Cybernetics,* **SMC-2**(2), (1972).

(*16*) F. L. Adelman and I. Adelman, Simulation of City Politics, presented at the Joint National Conference on Major Systems, Anaheim, Calif., Oct. 28, 1971.

Bibliography

Abt, C. B., Games for Learning, Chap. 3 in *Simulation Games in Learning,* S. S. Boocock and E. O. Schild, eds., Sage Publications, Inc., Beverly Hills, Calif., 1968.

Adelman, F. L., and I. Adelman, Simulation of City Politics, presented at the Joint National Conference on Major Systems, Anaheim, Calif., Oct. 28, 1971.

Adelman, H. M., System Analysis and Planning for Public Health Care in the City of New York, *Arch. Environ. Health,* **16** (Feb. 1968).

Adelson, M., The System Approach—A Perspective, *SDC Magazine,* **9** (10), (1966).

Alexander, C., *Notes on the Synthesis of Form,* Harvard University Press, Cambridge, Mass., 1964.

Alexander, T., Where Will We Put All That Garbage, *Fortune* (Oct. 1967).

Alonso, W., Aspects of Regional Planning and Theory in the United States, Working Paper—87, Institute of Urban and Regional Development, University of California, Berkeley, Calif.

Alonso, W., Industrial Location and Regional Policy in Economic Development, Working Paper 74 Institute of Urban and Regional Development, University of California, Berkeley, Calif., 1968.

Andrew, G. M., et al., University Student Registration Systems Design Using Simulation, *J. Educ. Data Processing.* 217–228 (1970).

Angel, S., Discouraging Crime Through City Planning, Working Paper 75, Institute of Urban and Regional Development, University of California, Berkeley, Calif., Feb. 1968.

Ansoff, H. I., and D. P. Slevin, An Appreciation of Industrial Dynamics, *Management Sci.* **14** (7), (1968).

Bartholomew, D. J., *Stochastic Models for Social Processes,* John Wiley & Sons, Inc., New York, 1967.

Besel, R., A Pupil Performance Measure for Criterion Referenced Instructional Programs, presented at the 40th National Meeting Operations Research Society of America, Anaheim, Calif., Oct. 27, 1971.

Blum, E., Urban Fire Protection: Studies of the Operations of the New York City Fire Department, Rept. R–681, Rand Institute, New York, Jan. 1971.

Blumstein, A., and R. Larson, Models of a Total Criminal Justice System, *Operations Research,* 199–232, Mar.-Apr. 1969.

Brown, A., and R. F. Kirby, Measuring Urban Performance, presented at the 40th National Meeting of the Operations Research Society of America, Oct. 27–29, 1971.

Bruno, J. E., The Function of Operations Research Specialists in Large Urban School Districts, *IEEE Trans. Systems Sci. Cybernetics,* **SSC-6** (4), (1970).

Campbell, H. S., et al., Alternative Development Strategies for Air Transportation in the New York Region 1970–1980, Rept. RM-5815-PA, The Rand Corporation, Santa Monica, Calif., Aug. 1969.

Carroll, S. J., A. H. Pascal, A Systems Analytic Approach to the Employment Problems of Disadvantage Youth, Rep. P-4045, Rand Corporation, Santa Monica, Calif., Mar. 1969.

Carter, G., and E. Ignall, Predicting the Actual Number of Fire-Fighting Units Dispatched, Rand Institute, New York, unpublished report.

Carter, G., and E. Ignall, A Simulation Model of Fire Department Operations: Design and Preliminary Results, Rept. R-632-NYC, Rand Institute, New York, Dec. 1970.

Chaiken, J., and R. Larson, Methods for Allocating Urban Emergency Units, Rept. R-680, Rand Institute, New York, May 1971.

Chaiken, J., and R. Larson, Methods for Allocating Urban Emergency Units: A Survey, Rept. P-4719, Rand Institute, New York, Oct. 1971.

Chaiken, J., and J. Rolph, Predicting the Demand for Fire Service, Rept. P-4625, Rand Institute, New York, May 1971.

Chamberlain, R. G., Crime Prediction Modeling, Rept. 650–126. Jet Propulsion Laboratory, California Institute of Technology, Pasadena, Calif., Apr. 23, 1971.

Churchman, C. W., *The Systems Approach,* Dell Publishing Company, Inc., New York, 1968.

Clinkscale, R. M., and M. P. Rymer, An Effectiveness Evaluation Model for Community-Based Social Service Delivery Systems, presented at the Joint National Conference on Major Systems, Anaheim, Calif., Oct. 29, 1971.

Colner, D., and D. Gilsinn, Fire Service Location—Allocation Models, Rept. 10833, National Bureau of Standards, Apr. 1972.

Conway, R., Operations Research and Hospital Design, presented to the 40th National ORSA Meeting, Oct. 28, 1971.

Cooper, G. R., and C. D. McGillem, *Methods of Signal and System Analysis,* Holt, Rinehart and Winston, Inc., New York, 1967.

Cordey, H. M., Retail Location Models, Working Paper 16, Centre for Environmental Studies, London.

Curry, D. A., A Systems Approach to Societal Problems, presented at the Joint National Conference on Major Systems, Anaheim, Calif., Oct. 28, 1971.

Dantzig, G. B., *Linear Programming and Extensions,* Princeton University Press, Princeton, N. J., 1963.

de Neufville, R., and J. Stafford, *Systems Analysis for Engineers and Managers,* McGraw-Hill Book Company, New York, 1971.

Donnelly, T. G., F. S. Chapin, Jr., and S. F. Weiss, A Probabilistic Model for Residential Growth, Institute for Research in Social Science, Chapel Hill, N. C., May, 1964.

Duke, R. D., Gaming Urban Systems, *Planning,* 293–300 (1965).

Dutton, J. M., and W. H. Starbuck eds., *Computer Simulation of Human Behavior,* John Wiley & Sons, Inc., New York, 1971.

Dyckman, J. W., Transportation in the Cities, *Sci. Am.,* **213** (3) (1965).

Elkin, R., The Systems Approach to Defining Welfare Programs, *Child Welfare,* **46** (2), 72–77 (1970).

Feldt, A. G., Operational Gaming in Planning Education, *Am. Inst. Planners J.* (Jan. 1966).

Final Report on the Development of a Criminal Court Calendar Scheduling Technique and Court Day Simulation, Programming Methods, Inc., New York, Mar. 1971.

Flagle, C. D., The Role of Simulation in the Health Services, *Am. J. Public Health,* **60** (12), (1970). 2386–2394.

Forrester, J. W., Counterintuitive Behavior of Social Systems, *Technol. Rev.,* **73** (3), (1971).

Forrester, J. W., *Industrial Dynamics,* The MIT Press, Cambridge, Mass., 1961.

Forrester, J. W., Industrial Dynamics—After the First Decade, *Management Sci.* **14** (7), (1968).

Forrester, J. W., *Urban Dynamics,* The MIT Press, Cambridge, Mass., 1969.

Friedman, J., Regional Planning as a Field of Study, *Am. Inst. Planners J.* (Aug. 1963).

Galvin, D. M., Child Care Centers—An Analysis Overview, presented at the 40th National Meeting of the Operations Research Society of America, Anaheim, Calif., Oct. 28, 1971.

Gass, S. I., *Linear Programming Methods and Applications,* McGraw-Hill Book Company, New York, 1958.

Goldman, T. A., *Cost-Effectiveness Analysis,* Praeger Publishers, Inc., New York, 1967.

Goldner, W., The Lowry Model Heritage, *Am. Inst. Planners J.,* (Mar., 1971).

Gordon, G., *System Simulation,* Prentice-Hall, Inc., Englewood Cliffs, N. J., 1969.

Gordon, G., and K. Zelin, A Simulation Study of Emergency Ambulance Service, *Trans. N. Y. Acad. Sci.,* Ser. II, **32** (4), 414–427 (1970).

Greenwood, P. W., Long Range Planning in the Criminal Justice System: What State Planning Agencies Can Do, Rept. P4379, Rand Corporation, Santa Monica, Calif., June 1970.

Gross, B. M., What Are Your Organization's Objectives: A General Systems Approach to Planning, *Human Relations, 195–217* (1965).

Hadley, G., *Linear Programming,* Addison-Wesley Publishing Company, Inc., Reading, Mass., 1962.

Hamilton, H. R., et al., *Systems Simulation for Regional Analysis: An Application to River Basin Planning,* The MIT Press, Cambridge, Mass., 1969.

Hare, V. C., Jr., *Systems Analysis: A Diagnostic Approach,* Harcourt, Brace, Jovanovich, Inc., New York, 1967.

Hartley, H. J., *Educational Planning–Programming–Budgeting, A Systems Approach,* Prentice-Hall, Inc., Englewood Cliffs, N. J., 1968.

Hartley, H. J., Limitations of Systems Analysis, *Phi Delta Kappan,* 515–519, May 1969.

Heller, N., and J. McEwen, The Use of an Incident Seriousness Index in the Deployment of Police Patrol Manpower, Rept. NI 71–036G, National Institute of Law Enforcement and Criminal Justice, Jan. 1972.

Heller, N., R. Markland, and J. Brockelmeyer, Partitioning of Police Districts into Optimal Patrol Beats Using a Political Districting Algorithm: Model Design and Validation, presented at the 40th National Meeting of the Operations Research Society of America, Anaheim, Calif.

Hill, D. M., A Growth Allocation Model for the Boston Region, *Am. Inst. Planners J.,* **31,** (2), (1965).

Hillier, F. S., and G. J. Kiekerman, *Introduction to Operations Research,* Holden-Day, Inc., San Francisco, 1967.

Hogg, J., The Siting of Fire Stations, *Operational Res. Quart.* **19** (3), 275–287 (1968).

Holbrook, R., M. King, and D. Webber, A Plan for the Development of a Laboratory of Criminal Justice, presented at the 40th National Meeting, Operations Research Society of America, Anaheim, Calif., Oct. 27, 1971.

Hoos, I. R., A Critical Review of Systems Analysis: The California Experience, Paper 89, Rec. 1968, Space Sciences Laboratory, University of California, Berkeley, Calif.

Horvath, W. J., The Systems Approach to the National Health Problem, *Management Sci.* **12** (10), B-391–395 (1966).

Howland, D., Toward a Community Health System Model, Chap. 9 in *Systems and Medical Care,* A. Sheldon et al., eds., The MIT Press, Cambridge, Mass., 1970.

IEEE Trans. Systems, Man. Cybernetics, **SMC-2** (2), (1972).

Irwin, N. A., Review of Existing Land-Use Forecasting Techniques, *Highway Rev. Board Record,* No. 88, 184–189.

Jacobs, J., *The Death and Life of Great American Cities,* Random House, Inc. (Vintage Books), New York, 1961.

Jennings, J. B., Blood Bank Inventory Control, Chap. 11 in *Analysis of Public Systems,* A. W. Drake, R. L. Keeney, and P. M. Morse, eds., The MIT Press, Cambridge, Mass., 1972.

Jennings, J. B., The Flow of Defendants Through the New York City Criminal Court in 1967, Rep. RM-6364-NYC, Rand Institute, New York, Sept. 1970.

Jennings, J. B., Quantitative Models of Criminal Courts, Rep. P-4641, Rand Institute, New York, May 1971.

Johnston, J., *Statistical Cost Analysis,* McGraw-Hill Book Company, New York, 1960.

Kadanoff, L. P., From Simulation Model to Public Policy, *Am. Scientist,* **60** (1), (1972).

Kakalik, J., and S. Wildhorn, Aids to Decision-Making in Police Patrol: An Overview of Study Findings, Rept. P-4614, The Rand Corporation, Santa Monica, Calif., Mar. 1971.

Kean, D. W., Humanistic Technology: A Contradiction in Terms, *The Humanist* (Jan.-Feb. 1972).

Kennedy, F. D., Development of a Community Health Service Simulation Model, *IEEE Trans. Systems Sci. Cybernetics,* **SSC-5** (3), 199–207 1969.

Kilbridge, M. D., R. P. O'Block, and P. V. Teplitz, A Conceptual Framework for Urban Planning Models, *Management Sci.,* **15** (6), 246–266 (1969).

Kleinmintz, B., *Personality Measurement,* Dorsey Press, Inc., Homewood, Ill., 1967.

Kolodney, S. E., and D. Daetz, Corrections Cost Projects, Rept. 458, Sylvania Electronic Systems, Mountain View, Calif., Jan. 1969.

Larson, R., Models for the Allocation of Urban Police Patrol Forces, MIT Operations Research Center Tech. Rept. 44, 1969.

Lowry, I. S., A Model of Metropolis, Rept. RM-4035-RC, Rand Corporation, Santa Monica, Calif., Aug. 1964.

Lowry, I. S., A Short Course in Model Design, *Am. Inst. Planners J.,* **31** (2), 160, 1965.

Luce, R. D., and H. Raiffa, *Games and Decisions,* John Wiley & Sons, Inc., New York, 1957.

McBride, C. C., Post Office Mail Processing Operations, in *Analysis of Public Systems,* A. W. Drake, R. L. Keeney, and P. M. Morse, eds., The MIT Press, Cambridge, Mass., 1972.

McCall, J. J., An Analysis of Poverty: A Suggested Methodology, Rept. RM–5739–OEO, Rand Corporation, Santa Monica, Calif., Oct. 1968.

McEwen, T., Allocation of Patrol Manpower Resources in the Saint Louis Police Department, Vols. I and II, 1966.

McLoughlin, J. B., Simulation for Beginners: The Planting of a Sub-regional Model System, *Regional Studies,* **3,** 313–323 (1969).

McLoughlin, J. B., *Urban and Regional Planning–A Systems Approach,* Praeger Publishers, Inc., New York, 1969.

McNanama, J., *Systems Analysis for Effective School Administration,* Parker Publishing Co., Inc., West Nyack, N. Y., 1971.

Maier, J. W., The Troubled City, Chap. 6 in *The Challenge to Systems Analysis,* G. J. Kelleher, ed., John Wiley & Sons, Inc., New York, 1970.

Manegold, R., and M. Silver, The Emergency Medical Care System, *J. Am. Med. Assoc.,* **200** (4), 124–218 (1967).

Meier, R. L., and R. D. Duke, Gaming Simulation for Urban Planning, *Amer. Inst. Planners J.,* (Jan. 1966).

Merrill, F., and L. Schrage, Efficient Use of Jurors: A Field Study and Simulation Model of a Court System, *Washington University Law Quarterly,* No. 2, 151–183 (1969).

Millward, R. E., PPBS: Problems of Implementation, *Am. Inst. Planners J.,* (Mar. 1968).

Morse, P. M., Library Models, Chap. 12 in *Analysis of Public Systems,* A. W. Drake, R. L. Keeney and P. M. Morse, eds., The MIT Press, Cambridge, Mass., 1972.

National Industrial Conference Board and the Opinion Research Corporation, Perspectives for the '70s and '80s, an experimental forecast conducted in 1968, 1970.

Navarro, J., and J. Taylor, Data Analyses and Simulation of the Court System in the District of Columbia for the Processing of Felony Defendants, *Task Force Report: Science and Technology,* President's Commission on Law Enforcement and Administration of Justice, Washington, D. C., 1967, 37–44, 199–215.

Nilsson, E., and J. Swartz, Jr., Application of Systems Analysis to the Alexandria, Virginia, Fire Department, Rept. 10454, National Bureau of Standards, Feb. 1972.

Nobile, P., ed., *The Con III Controversy,* Pocket Books, Inc., New York, 1971.

North, D. W., A Tutorial Introduction to Decision Theory, *IEEE Trans. Systems Sci. Cybernetics, SSSC-4* (3), (1968).

Optner, S. L., *Systems Analysis for Business Management,* 2nd ed., Prentice Hall, Inc., Englewood Cliffs, N. J., 1968.

Orlando, J. A., and A. J. Pennington, Build—A Community Development Simulation Game, taken from a manuscript prepared for the 36th National Meeting, Operations Research Society of America, Miami Beach, Fla., Nov. 10–12, 1969.

Owen, J., Emergency Services Must Be Reorganized, *The Modern Hospital,* 84–90 (Dec. 1966).

Page, D. A., The Federal Planning–Programming–Budgeting System, *Am. Inst. Planners J.,* 256–259 (July 1967.)

Pardee, F. S., et al., Measurement and Evaluation of Transportation System Effectiveness, Rept. RM–5869–DOT, Rand Corporation, Santa Monica, Calif., Sept. 1969.

Pascal, A. H., New Departures in Social Services, Rept. P–4079, Rand Corporation, Santa Monica, Calif., Apr. 1969.

Prabho, N., *Queues and Inventories, A Study of Their Basic Stochastic Processes,* John Wiley & Sons, Inc., New York, 1965.

Prest, A. R., and R. Turvey, Cost-Benefit Analysis: A Survey, *Econ. J.* No. 300 (Dec. 1965).

Revelle, C., D. Marks, and J. C. Liebman, An Analysis of Private and Public Sector Location Models, *Management Sci.,* **16** (11), (1970).

Rockett, J., Objectives and Pitfalls in the Simulation of Building Fires with a Computer, *Fire Technol.,* **5** (4), 311–322 (1969).

Rogers, A., *Matrix Analysis of Inter-regional Population Growth and Distribution,* University of California Press, Berkeley, Calif., 1968.

Savas, E. S., Simulation and Cost-Effectiveness Analysis of New York's Emergency Ambulance Service, *Management Sci.,* **15** (12), B–608–618 (1969).

Schlager, K. J., A Land Use Plan Design Model, *Am. Inst. Planners J.,* **31** (2), (1965).

Seiler, K., III, *Introduction to Systems Cost-Effectiveness,* John Wiley & Sons, Inc., New York, 1969.

Shauelson, R. J., and M. R. Munger, Individualized Instruction: A Systems Approach, *J. Educ. Res.* **63** (6), (1970).

Shubik, M., Gaming: Costs and Facilities, *Management Sci.* **14** (11), 629–660 (1968).

Siegel, S., and L. E. Fouraker, *Bargaining and Group Decision Making: Experiments in Bilateral Monopoly,* The Macmillan Company, New York, 1960.

Simonnard, M., *Linear Programming,* Prentice-Hall, Inc., Englewood Cliffs, N. J., 1966.

Spindler, A., Decision Making: The Target of Systems Science in Social Programs, presented at the Joint National Conference on Major Systems, Anaheim, Calif., Oct. 29, 1971.

Steger, W. A., The Pittsburgh Urban Renewal Simulation Model, *Am. Inst. Planners J.,* **31** (2), (1965).

Stonebraker, M., A Simplification of Forrester's Model of an Urban Area, *IEEE Trans Systems, Man, Cybernetics,* **SMC-2** (4), (1972).

Suits, D. B., Forecasting and Analysis with an Econometric Model, *Am. Econ. Rev.,* 104–132 (1962).

Teitz, M. B., Cost Effectiveness: A Systems Approach to Analysis of Urban Services, *Amer. Inst. of Planners J.,* (Sept. 1968).

Tenzer, A., J. Benton, and C. Teng, Applying the Concepts of Program Budgeting to the New York City Police Department, Rept. RM–5846–NYC, Rand Corporation, Santa Monica, Calif., June 1969.

Thompson, R. B., *A Systems Approach to Instruction,* The Shoe String Press, Hamden, Conn., 1971.

Thorne, E. M., Regional Input–Output Analysis, in *Regional and Urban Studies,* S. C. Orr and J. B. Cullingworth, eds., Sage Publications, Beverly Hills, Calif.

Twelker, P. A., Designing Simulation Systems, *Educ. Technol.,* 64–70 (Oct. 1969).

Twelker, P. A., ed., *Instructional Simulation Systems,* Continuing Education Publications, Corvallis, Ore., 1969.

U. S. Department of Health, Education, and Welfare, Toward a Social Report, Government Printing Office, Washington, D. C., 1969.

Van Arsdol, M. D., Jr., Metropolitan Growth and Environmental Hazards: An Illustrative Case, *Ekistics,* 48–50 (1966).

VanDueseldorp, R. A., et al., *Educational Decision-Making Through Operations Research,* Allyn and Bacon, Inc., Boston, 1971.

Vidale, R. F., University Programs in Systems Engineering, *IEEE Trans. Systems Sci. Cybernetics,* **SSC-6** (3), 217–228 (1970).

von Neumann, J., and O. Morgenstern, *Theory of Games and Economic Behavior,* 2nd ed., Princeton University Press, Princeton, N. J., 1947.

Webber, M., Systems Planning for Social Policy, taken from *Readings in Community Organization Practice,* R. M. Kramer, et al., eds., Prentice-Hall, Inc. Englewood Cliffs, N.J., 1969.

Weitz, H., A Model for the Simulation of the Fire Services of an Urban Community, *Fire J.* 48–55 (Jan. 1969).

Wilde, D. J., and C. S. Beightler, *Foundations of Optimization,* Prentice-Hall, Inc., Englewood Cliffs, N. J., 1967.

Wilson, A. G., Models in Urban Planning: A Synoptic Review of Recent Literature, *Urban Studies,* **5** (3), 249–276 1968.

Wilson, A. G., Research for Regional Planning, *Regional Studies,* **3,** 3–14 (1969).

Wilson, O. W., *Police Administration,* 2nd ed., McGraw-Hill Book Company, New York, 1963.

Wohl, M., and B. V. Martin, *Traffic System Analysis for Engineers and Planners,* McGraw-Hill Book Company, New York, 1967.

Wolfe, H. B., and M. L. Ernst, Simulation Models and Urban Planning, in *Operations Research for Public Systems,* P. M. Morse, ed., The MIT Press, Cambridge, Mass., 1967.

Wolfgang, M., and T. Sellin, *The Measurement of Delinquency,* John Wiley & Sons, Inc., New York, 1964.

Wong, A., and T. Au, A Dynamic Model for Planning Patient Care in Hospitals, *IEEE Trans. Systems Sci. Cybernetics,* **SSC-5** (3), 199–207 (1969).

Wormeli, P., and S. E. Kolodney, Transaction Based Statistics for Criminal Justice Management Decision Making, presented at the 40th National Meeting, Operations Research Society of America, Anaheim, Calif., Oct. 27, 1971.

Young, J. W., W. F. Arnold, and J. W. Brewer, Parameter Identification and Dynamic Models of Socioeconomic Phenomena, *IEEE Trans. Systems, Man. Cybernetics,* **SMC-2** (4), (1972).

Index